# Gene Expression

*Edited by Fumiaki Uchiumi*

Published in London, United Kingdom

Gene Expression
http://dx.doi.org/10.5772/intechopen.98147
Edited by Fumiaki Uchiumi

Contributors
Kalyani M. Barbadikar, Nakul D. Magar, Priya Shah, K. Harish, Tejas C. Bosamia, Maganti Sheshu Madhav, Raman Meenakshi Sundaram, Chirravuri Naga Neeraja, Satendra Kumar Mangrauthia, Amol Phule, Yogesh M. Shukla, Harshvardhan N. Zala, Parvaiz Yousuf, Preeti Dabas, Ja Yil Lee, Na Young Cheon, Subin Kim, Uma Siangphoe, Pranjal Kumar, Nikita Bhandari, Jenny Paredes, Jennie Williams, Shrey Thaker, Fumiaki Uchiumi, Tjahjadi Robert Tedjasaputra, Mochammad Hatta, Muhammad Nasrum Massi, Agussalim Bukhari, Rosdiana Natzir, Ilhamjaya Patellongi, Muhammad Lutfi Parewangi, Prihantono, Marcellus Simadibrata, Rinda Nariswati, Shirly Elisa Tedjasaputra, Vincent Tedjasaputra, Rina Masadah, Jonathan Salim, Andi Asadul Islam



First published in London, United Kingdom, 2022 by IntechOpen
IntechOpen is the global imprint of INTECHOPEN LIMITED, registered in England and Wales, registration number: 11086078, 5 Princes Gate Court, London, SW7 2QJ, United Kingdom

British Library Cataloguing-in-Publication Data
A catalogue record for this book is available from the British Library

Additional hard and PDF copies can be obtained from orders@intechopen.com

Gene Expression
Edited by Fumiaki Uchiumi
p. cm.
Print ISBN 978-1-80355-621-5
Online ISBN 978-1-80355-622-2
eBook (PDF) ISBN 978-1-80355-623-9

For EU product safety concerns:
IN TECH d.o.o., Prolaz Marije Krucifikse Kozulić 3, 51000 Rijeka, Croatia,
info@intechopen.com or visit our website at intechopen.com.

# Meet the editor

Fumiaki Uchiumi, Professor of Pharmaceutical Sciences, Tokyo University of Science, grew up in Japan and received his bachelor's degree in Chemistry from Tokyo University of Science in 1987. In 1993, after obtaining his Ph.D. in Molecular Biology from Tokyo University, he joined Professor S. Tanuma's Laboratory at Tokyo University of Science as an assistant professor. He obtained his second Ph.D. in Pharmaceutical Science from Tokyo University of Science in 1999, and in 2000 he was promoted to the position of lecturer at the same university. He then went abroad as a post-doctoral researcher in the United States–Japan Cooperative Cancer Research Program at Professor E. Fanning's Laboratory, Vanderbilt University, USA 2000–2001. Dr. Uchiumi was promoted to associate professor and then full professor at Tokyo University of Science in 2010 and 2016, respectively.

# Contents

# Preface

It is believed that the first ancestral organism on Earth was created from the "RNA world." It could be said that all organisms, including humans, are still living in the RNA world. The Central Dogma of molecular biology is the flow of genetic information from genomic DNA to various proteins. The process is mediated by (classical) non-coding RNAs, including mRNAs, tRNAs, and rRNAs. Genomic DNA needs to be replicated just before a cell divides. Alternatively, RNAs are always synthesized and degraded if the cell is living. Although DNA nucleotides could be synthesized from RNA nucleotides, they need to be processed through complicated redox reactions. Surely, DNA is an essential material for storing genetic information, but RNA is a functional molecule. We know that gene expression is regulated by multiple steps, namely transcription, RNA processing, and translation, which should be considered and discussed if we precisely understand gene expression. Before the discussion, we need reliable systems to detect and characterize DNA and RNA. Fortunately, at present, whole human genomic sequences have been revealed, and just a single molecule of specific RNA is enough for sequencing. This book provides a comprehensive overview of gene expression and its role in human disease. It includes nine chapters organized into two sections.

Section 1 reviews the developments in gene expression analyses that revealed molecular mechanisms in controlling gene expression. Chapter 1 discusses some of the current concepts for the precise understanding of gene expression. It was recently revealed that gene expression can be regulated by loop structures of chromosomes or DNA/RNA hybrid loops. Transcription factor-binding sequences as well as epigenetic regulation, which affect chromatin structure, play essential roles in gene expression. Naturally, it should depend on nutrients and metabolites. Chapter 2 by Dabas discusses protein factors that regulate the transcription elongation process in eukaryotes. Elucidating the molecular mechanisms sheds light on the development of therapeutics for transcription-associated human diseases. Chapter 3 by Cheon et al. highlights recent progress in the analyses of the R-loop, which is a specific DNA–RNA hybrid structure. It also discusses multiple biologically significant functions. Chapter 4 by Kumar et al. describes lncRNAs as gene expression controlling factors. The functions of lncRNAs and miRNAs should be examined to understand complicated gene expression systems in eukaryotic cells. Chapter 5 by Magar et al. discusses the history, present clinical applications, and perspectives of gene expression analyses.

Section 2 examines the relationships between transcriptional control and human disease. It also discusses the possibilities of DNA/RNA analyses for diagnostics and the application of DNA/RNA drugs for treatment. Chapter 6 by Tedjasaputra et al. highlights the importance of analyzing APC and MSH2 gene expression in human colorectal cancer. The earlier cancer can be identified, the better. Chapter 7 by Yousuf et al. discusses the identification of circular DNAs and their functions in cancer generation. Chapter 8 by Williams et al. discusses aberrant gene expression in cancer. It is suggested that analyzing epigenetic regulation and miRNAs is important for predicting and developing treatments for colon cancers. Finally, Chapter 9 by Siangphoe et al. presents a Bayesian random-effects meta-analysis of gene expression in Alzheimer's disease patients.

If it is discovered that some diseases are dependent on the expression of specific genes, these diseases could be treated by a drug(s) that up-/down-regulate the expression of those genes. It has been shown that small RNAs or miRNAs can affect transcription and translation. Thus, nucleotide-based drugs are expected to be developed. More research in this field will play a key role in the development of new therapeutics based on molecular biology.

During the time of editing this book, Dr. Kumao Toyoshima, a pioneer in the study of oncogenes and proto-oncogenes, passed away. I dedicate this book to his great achievements, and I would like to acknowledge all his suggestions and advice that encouraged us to study molecular biology.

**Fumiaki Uchiumi, Ph.D.**
Professor,
Department of Gene Regulation,
Tokyo University of Science,
Tokyo, Japan

## Section 1

# Molecular Mechanisms and Analyses of Gene Expression

Chapter 1

# Introductory Chapter: Gene Expression in Eukaryotic Cells

*Fumiaki Uchiumi*

## 1. Introduction

"Central Dogma" explains how information from genes to proteins flows. Genes should be transcribed into messenger ribonucleic acids (mRNAs) in nuclei, then they are processed and delivered to the cytoplasm where they are translated into polypeptides (proteins). We, molecular biologists, know that gene expression in mammalian cells is controlled at multiple stages. First, gene expression is epigenetically regulated by chromatin structures, which depend on deoxyribonucleic acid (DNA) methylation and histone modifications. Then, transcription initiation, elongation, and termination occur, and RNA could be maturated in nuclei. Additionally, non-coding RNAs (ncRNAs), including miRNAs, affect gene expression. Moreover, loop structures of DNAs also play roles in gene expression. One of the recent striking topics is the identification of extrachromosomal circular DNAs (eccDNAs), and R-loop formation that is mediated by the interaction between DNAs and RNAs. In summary, the regulation of gene expression is a very much complicated system. In this chapter, I would review how gene expression controlling systems in mammalian cells are presently understood. I hope that we would be inspired to think over essential problems to be dissolved toward the progress of medical sciences.

## 2. Gene expression, which is regulated by multiple steps

It has been widely known how general transcription factors (GTFs) execute RNA pol II dependent transcription in mammalian cells [1]. I have ever reviewed transcription control systems in mammalian cells, especially focusing on the possibility of the application of transcription-regulating mechanisms on gene therapy [2, 3]. This time, we would challenge the most fundamental problems that should be addressed before discussing the practical use of the transcription system. Generally, gene expression is defined as the producing rate of the mature mRNAs that are to be utilized for the translation of polypeptides. Transcription begins with initiation, elongation, and ends with the termination process. In eukaryotes, the premature RNAs are to be processed by splicing [4], 5′-cap modification [5, 6], and polyadenylation or poly A tail addition [7]. The matured RNAs are transported to the cytoplasm to be used for translation. RNAs are unstable and they are easily degraded by ribonucleases (RNases), which are ubiquitously present in all kinds of cells. In human, it has been known that at least 13 belong to hRNase A superfamily [8]. Some of them play roles in the host defense system [9], and others are required for host and mitochondrial

DNA replication [10–12]. Thus, as same as other biologically synthesized polymers, amounts of matured RNAs are appropriately regulated by the balance between synthesis and degradation.

## 3. Loop Structures that regulate transcription

Generally, it is thought that gene expression is considerably regulated by the initiation step that is dependent on sequence-specific TFs [2, 3]. The loop structure formation by the interaction between enhancer and core promoter [13–16] is thought to be essential for GTFs to start transcription in the right direction. The enhancer-promoter loop can be made by two double-stranded DNAs (dsDNA) and the most 5′-upstream RNA strand. The formation of chromatin loop clusters may be mediated by specific proteins, including CCCTC-binding factor (CTCF) [17]. The loop formation between dsDNAs might be associated with chromosome-wide spreading gene-silencing mechanism as that has been well studied for *Xist* [18]. The possible molecular model is that RNA strands may play a role as bridges between DNAs. On the other hand, R-loop structure, which enables the recycling of TFs and RNA pol II, is made by communication between transcription initiation and termination sites [19–22]. The R-loop, which is composed of separated complementary DNA strands or "transcription bubble" and elongating RNA, could induce antisense transcription at enhancer elements, and transcription initiation and termination sites [23]. Thus, the DNA-RNA triplet R-loop can also play an essential role in the bidirectional transcription on eukaryotic chromosomes. The R-loop resembles to the replication bubble, which is made by separated complementary strands and a short RNA that is required for the leading strand synthesis. Therefore, the generation of R-loops needs to be accurately controlled. If the dysregulation occurred, it may lead to conflicts between transcription and replication machinery, which will cause DNA damage and cell death. The problem might be dissolved by several key regulators, including ATR, CHEK1, and BRD4 [24, 25]. The promoter regions of eukaryotic genes have been mainly studied by the reporter assay system with plasmid vectors that express such as luciferase. However, usually, it has not been paid attention to the 3′-untranslated regions (3′-UTRs). It has been presumed that 3′-UTRs play important roles in the regulation of stability, localization of transcripts, and translation [26–28]. If the R-loop has a considerable effect on the production of immature transcripts, it should not be ignored.

## 4. Transcriptional direction might be epigenetically regulated

Epigenetic regulation is mainly executed by DNA methylation and histone modification [29] that control chromatin structure to regulate genomic imprinting [30] and cellular senescence [31]. The direction of transcription might be dependent on DNA methylation, which is regulated by DNA methyltransferases (DNMTs) and the Ten-eleven-translocation (TET) enzymes [29, 32, 33]. CpG islands, which can be a methylation target [34], are commonly present at bidirectional transcription loci in human chromosomes [35, 36]. The bidirectional promoter regions have more GC-rich sequences but less TATA boxes than unidirectional promoters [37]. The majority (>80%) of CpGs in the human genome of the somatic cells are methylated, apart from actively transcribed regions, including promoters and enhancers [38]. However, because whole-genome methylome analyses identified differentially methylated regions (DMRs) in the human chromosomes [39], not all GC-rich sequences are the targets for methylation. Notably, specific TFs preferentially bind to the methylated CpGs [40]. ETS family protein PU.1 (SPI1) forms a complex with Dnmt3a/b to bring site-specific methylation to cause down-regulation of transcription [41]. The GC-box recognizing Sp1 can interact with DNMT1 in human cells [42]. Moreover, C/EBPα, Klf4, and Tfcp2l1 can affect Tet2 to demethylate specific promoters to induce pluripotency of cells [43]. In summary, site-specific DNA methylation/demethylation, modulating affinities with specific TFs, might determine which unidirectional or bidirectional transcription would be preferred.

## 5. Nutrients and metabolites dependent gene expression

Prokaryotic lactose operon system has been generally known. Nutrients or metabolites must be monitored to control transcription adequately in mammalian cells [44]. Glucose regulates the transcription of genes that encode lipogenesis-associated proteins through activation of the LXR (NR1H) factors [45]. HDL metabolism functioning protein-encoding genes are induced by glucose [46]. A glucose derivative molecule, 2-deoxy-D-glucose (2-DG) upregulates promoter activities of the *TERT* and *WRN* genes in HeLa S3 cells [47].

Fatty acids can affect transcription mediated by PPARs [48], SREBP-1 [49], and other TFs [50]. The n-butyrate (butyric acid), which is produced by gut bacteria, or sodium butyrate regulates gene expression in human cells [51]. The transcriptional regulation by butyrate has been explained by an inhibitory action on histone deacetylase [52] and increasing stabilities of mRNAs [53]. Notably, Sp1 that recognizes a GC-rich sequence [54] is hyper-acetylated by butyrate in human colon cells [55]. Therefore, transcription of specific genes, if their promoter contains a Sp1 binding element(s), could be affected by a lipophilic acid, which stops proliferation [56]. Other TFs, including ETS family ETV1 (ER81) [57], ETV4 (PEA3) [58], and p53 [59], are also activated by butyrate-induced signals.

Amino acids also regulate transcription. For example, glutamine responsive genes have been identified [60]. Leucine starvation induces promoter activity of the *CHOP* gene, encoding the CCAAT/enhancer binding protein, through binding of ATF-2 to an amino acid response element (AARE) [61]. Some amino acids are to be metabolized to acetyl-CoA and a methyl-group donor S-adenosylmethionine. They are the substrates for histone acetylation and DNA methylation, respectively. Tryptophan could be metabolized to a nicotinamide adenine dinucleotide ($NAD^+$), which is also produced from niacin (nicotinic acid and nicotine amide) or vitamin $B_3$, is not only required for mitochondrial functions, but also for poly (ADP-ribosyl) ation that is catalyzed by poly(ADP-ribose) polymerases (PARPs) [62, 63]. It has been suggested that $NAD^+$ regulates transcription in mammalian cells [64, 65]. In mice, the hepatic and cytosolic NADH/$NAD^+$ ratio alters the circulating α-hydroxybutyrate level, which could be an early biomarker for diabetes [66, 67]. Decrease in $NAD^+$ concentration has been suggested to be associated with aging or age-related diseases [68]. Notably, 2-DG and *trans*-resveratrol, which up-regulate $NAD^+$/NADH ratio in HeLa S3 cells [69], can activate the human *WRN* and *TERT* promoters [47, 70].

## 6. Euchromatin and heterochromatin

Euchromatin and heterochromatin represent specific structures of eukaryotic chromosomes, which are transcriptionally active and inactive, respectively [71]. Generally, heterochromatin represents a state where chromosomes are attached to a nuclear membrane with nuclear lamina [72]. The formation of the heterochromatin is thought to be under epigenetic regulation, during both the development and aging processes of mammalian cells. Heterochromatin does not only affect transcription but also protects chromosomes from mechanical stresses [73, 74]. Relationships between heterochromatin and DNA-repair systems have been proposed [75]. The DNA-repair systems could dominantly work on the euchromatin, and that is enabled by the stabilization of chromosomes by heterochromatin structure. It might be controversial that DNA repair factors cause heterochromatinization, but PARP-1 and 2, which play roles in the DNA-repair system, can contribute to the maintenance of heterochromatin [76]. The telomeric region where PARP or tankyrase is located plays a role in the maintenance of the ends of chromosomes [77]. The shortening of telomeres may lead to severe chromosomal instability that accelerates cellular senescence and cancer generation [78], suggesting that PARP enzymes protect telomeres by heterochromatinization. Moreover, the PARP modulates chromatin structure when it functions at centromeres [79]. Overall, PARP-dependent DNA repair systems are not only required for the conservation of nucleotide sequences of functional proteins but also for the maintenance of chromosomal structures that are constructed by specific or repetitive sequences. That would partly explain the reason why telomeres and centromeres might have excluded translocations of protein-encoding genes and transposons. This might also explain why PARP-1 does not have a preference for specific DNA sequences, surely it can find DNA breakage to load poly(ADP-ribose) to chromatin-associating proteins, including histones [80] and p53 [81]. The introduction of poly(ADP-ribose) on TFs may suppress transcription [82] and the repair system will work well in the promoter regions. Activation of PARPs consumes $NAD^+$ to synthesize poly(ADP-ribose), which is required for indicating the DNA damaging sites. That will cause a reduction in $NAD^+$-dependent transcription of mitochondrial protein-encoding genes [65]. Taken together, poly(ADP-ribosyl)ation plays a role in keeping a balance between DNA-repair, energy production, and transcription, maintaining chromosomal structures.

## 7. Concluding remarks

"Gene Expression" can be discussed from many points of view. Because it includes many biological events that are executed by various proteins and RNAs. Although I have not reviewed the recent progress in studies of non-coding RNAs, they play essential roles in transcriptional regulation [83]. "Gene Expression" is regulated by many stresses that can modulate DNA structures, loop formations, and epigenetic states. Dysregulation of "Gene Expression" will cause aging-related diseases. Hopefully, artificial transcription controlling systems will be developed and clinically applied to cancer and neurodegenerative diseases in the future. Nevertheless, we have not yet reached a conclusion or even a hypothesis on how gene expression system has been established and how it developed through a long evolution process. All organisms, including prokaryotes, archaea, and prokaryotes, would not live without accurate execution of the transcription system. The exceptions are viruses that just utilize the infected host cell system. Among them, some retroviruses are unique in carrying oncogenes to cause cancer and lymphoma [84]. Their genes are RNAs to be reverse transcribed to DNAs, which can be integrated into host cell chromosomes. The composition of the genome is characteristic, having long terminal repeats (LTRs). Interestingly, many retrovirus-like elements or transposons, including LINEs and SINEs, have been suggested to regulate gene expression, by both transcriptional and post-transcriptional mechanisms [85]. Every protein-encoding gene has a transcription start and termination site. How have genes acquired promoters and terminators, which are present on the 5′-upstreams and 3′-downstreams, respectively? The loop structures and extrachromosomal circular DNAs (eccDNAs), which are frequently identified when DNA amplification occurs, might give us a hint [86].

The dsDNA loops are thought to be formed when DNA damage was induced [87] or when a rearrangement of genes occurs in immune cells [88]. Although it is a hypothesis, the generation of multiple DNA replication initiation sites in eukaryotic chromosomes suggests that linear chromosomes might have been evolved from the fusion of multiple circular chromosomes (**Figure 1**). To prove this hypothesis, hot spot junction sites, which are the same as tentative dsDNA break sites, should be identified (**Figure 1C**). Loop or circular DNAs are formed at the time when dsDNAs integrated in or released from chromosomes. Therefore, there are both chances to gain or lose DNAs. If it occurred at chromosomal crossover during meiosis, the acquired or lost DNA sequences would be inherited to descendants. Elucidation of the biological meanings of transposition and amplification of genes will answer the question of how genes acquired promoters and terminators through evolution.

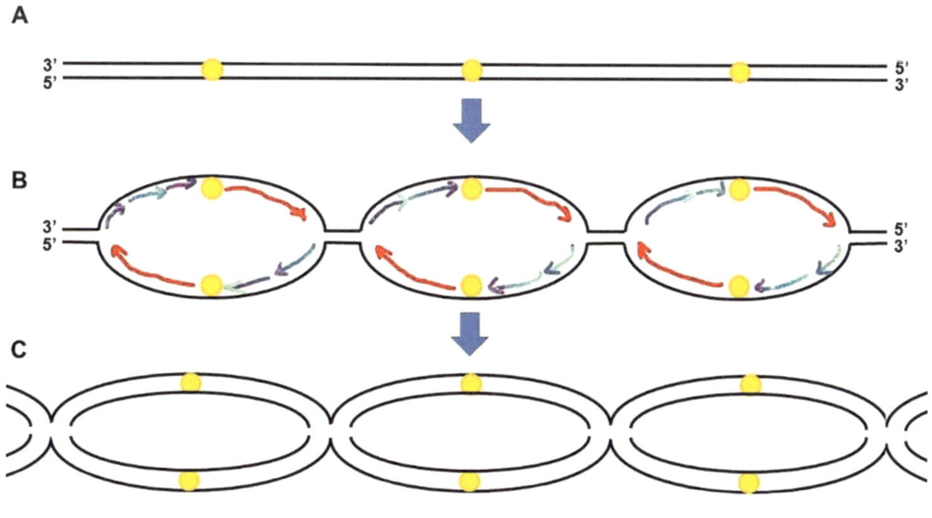


**Figure 1.**
*DNA replication of eukaryotic cells. (A) Multiple DNA replication initiation sites (yellow circles) are distributed all over chromosomes. (B) Leading strands (red) and lagging strands (blue) are synthesized from the initiation sites when S-phase started. (C) At the sites where opposite direction proceeding leading strands collide, the temporary generated dsDNA breaks would be generated to be ligated. Taken together, multiple circular DNAs, which are to be ligated with, might be generated at the end of the S-phase.*

## Author details

Fumiaki Uchiumi
Faculty of Pharmaceutical Sciences, Department of Gene Regulation, Tokyo University of Science, Noda, Chiba-ken, Japan

*Address all correspondence to: f_uchiumi@rs.tus.ac.jp

## References

[1] Carey MF, Peterson CL, Smale ST. A primer on transcriptional regulation in mammalian cells. In: Carey MF, Peterson CL, Smale ST, editors. Transcriptional Regulation in Eukaryotes: Concepts, Strategies, and Techniques. 2nd ed. Cold Spring Harbor, NY: Cold Spring Harbor Laboratory Press; 2009. pp. 1-45

[2] Uchiumi F. Current studies in transcriptional control system; toward the establishment of therapies against human diseases. In: Uchiumi F, editor. Gene Expression and Regulation in Mammalian Cells. London, UK: InTechOPEN; 2018. pp. 3-13

[3] Uchiumi F, Asai M. Gene expression controlling system and its application to medical sciences. In: Uchiumi F, editor. Gene Expression and Control. London, UK: InTechOPEN; 2019. pp. 3-10

[4] Yan C, Wan R, Shi Y. Molecular mechanisms of pre-mRNA splicing through structural biology of the spliceosome. Cold Spring Harbor Perspectives in Biology. 2019;**11**(1):a032409

[5] Muthukrishnan S, Both GW, Furuichi Y, Shatkin AJ. 5′-terminal 7-methylguanosine in eukaryotic mRNA is required for translation. Nature. 1975;**255**(5503):33-37

[6] Sikorski PJ, Warminski M, Kubacka D, Ratajczak T, Nowis D, et al. The identity and methylation status of the first transcribed nucleotide in eukaryotic mRNA 5′ cap modulates protein expression in living cells. Nucleic Acids Research. 2020;**48**(4):1607-1626

[7] Stewart M. Polyadenylation and nuclear export of mRNAs. The Journal of Biological Chemistry. 2019;**294**(9):2977-2987

[8] Lee HH, Wang YN, Hung MC. Functional roles of the human ribonuclease A super family in RNA metabolism and membrane receptor biology. Molecular Aspects of Medicine. 2019;**70**:106-116

[9] Rosenberg HF. RNase A ribonucleases and host defense: An evolving story. Journal of Leukocyte Biology. 2008;**83**(5):1079-1087

[10] Parajuli S, Teasley DC, Murali B, Jackson J, Vindigni A, et al. Human ribonuclease H1 resolves R-loops and thereby enables progression of the DNA replication fork. The Journal of Biological Chemistry. 2017;**292**(37):15216-15226

[11] Mentegari E, Crespan E, Bavagnoli L, Kissova M, Bertoletti F, et al. Ribonucleotide incorporation by human DNA polymerase eta impacts translesion synthesis and RNase H2 activity. Nucleic Acids Research. 2017;**45**(5):2600-2614

[12] Posse V, Al-Behadili A, Uhler JP, Clausen AR, Reyes A, et al. RNase H1 directs origin-specific initiation of DNA replication in human mitochondria. PLoS Genetics. 2019;**15**(1):e1007781

[13] van Arensbergen J, van Steensel B, Bussemaker HJ. In search of the determinants of enhancer-promoter interaction specificity. Trends in Cell Biology. 2014;**24**(11):695-702

[14] Mora A, Sandve GK, Gabrielsen OS, Eskeland R. In the loop: Promoter-enhancer interactions and bioinformatics. Briefings in Bioinformatics. 2016;**17**(6):980-995

[15] Robson MI, Ringel AR, Mundlos S. Regulatory landscaping: How enhancer-promoter communication is sculpted in 3D. Molecular Cell. 2019;**74**(6):1110-1122

[16] Yokoshi M, Segawa K, Fukaya T. Visualizing the role of boundary elements in enhancer-promoter communication. Molecular Cell. 2020;**78**(2):224-235.e5

[17] Li Y, Haarhuis JHI, Sedeño Cacciatore Á, Oldenkamp R, van Ruiten MS, et al. The structural basis for cohesin-CTCF-anchored loops. Nature. 2020;**578**(7795):472-476

[18] Yamada N, Ogawa Y. Mechanisms of long noncoding Xist RNA-medated chromosome-wide gen silencing in X-chromosome inactivation. In: Kurokawa R, editor. Long Noncoding RNAs. Dordrecht Heidelberg London New York: Springer Science+Business Media; 2015. pp. 151-171

[19] Martin M, Cho J, Cesare AJ, Griffith JD, Attardi G. Termination factor-mediated DNA loop between termination and initiation sites drives mitochondrial rRNA synthesis. Cell. 2005;**123**(7):1227-1240

[20] El Kaderi B, Medler S, Raghunayakula S, Ansari A. Gene looping is conferred by activator-dependent interaction of transcription initiation and termination machineries. The Journal of Biological Chemistry. 2009;**284**(37):25015-25025

[21] Bratkowski M, Unarta IC, Zhu L, Shubbar M, Huang X, et al. Structural dissection of an interaction between transcription initiation and termination factors implicated in promoter-terminator cross-talk. The Journal of Biological Chemistry. 2018;**293**(5):1651-1665

[22] Malig M, Hartono SR, Giafaglione JM, Sanz LA, Chedin F. Ultra-deep coverage single-molecule R-loop footprinting reveals principles of R-loop formation. Journal of Molecular Biology. 2020;**432**(87):2271-2288

[23] Tan-Wong SM, Dhir S, Proudfoot NJ. R-loop promote antisense transcription across the mammalian genome. Molecular Cell. 2019;**76**(4):600-616

[24] Matos DA, Zhang JM, Ouyang J, Nguyen HD, Genois M, et al. ATR protects the genome against R loops through a MUS81-triggered feedback loop. Molecular Cell. 2019;**77**(3):514-527

[25] Edwards DS, Maganti R, Tanksley JP, Luo J, Park JJH, et al. BRD4 prevents R-loop formation and transcription-replication conflicts by ensuring efficient transcription elongation. Cell Reports. 2020;**32**(12):108166

[26] Geissler R, Grimson A. A position-specific 3′ UTR sequence that accelerates mRNA decay. RNA Biology. 2016;**13**(11):1075-1077

[27] Mayr C. Regulation by 3′-untranslated regions. Annual Review of Genetics. 2017;**51**:171-194

[28] Mayr C. What are 3′ UTRs doing? Cold Spring Harbor Perspectives in Biology. 2019;**11**(19):a034728

[29] Loaeza-Loaeza J, Beltran AS, Hernández-Sotelo D. DNMTs and impact of CpG content, transcription factors, consensus motifs, lncRNAs, and histone marks on DNA methylation. Genes (Basel). 2020;**11**(11):1336

[30] Meng H, Cao Y, Qin J, Song X, Zhang Q, et al. DNA methylation, its mediators and genome integrity. International Journal of Biological Sciences. 2015;**11**(5):604-617

[31] Yang N, Sen P. The senescent cell epigenome. Aging (Albany NY). 2018;**10**(11):3590-3609

[32] Ismail JN, Ghannam M, Al Outa A, Frey F, Shirinian M. Ten-eleven translocation proteins and their role beyond DNA demethylation—what we can learn from the fly. Epigenetics. 2020;**15**(11):1139-1150

[33] Li E, Zhang Y. DNA methylation in mammals. Cold Spring Harbor Perspectives in Biology. 2014;**6**(5):a019133

[34] Lakshminarasimhan R, Liang G. The role of DNA methylation in cancer. In: Jeltsch A, Jurkowska RZ, editors. DNA Methyltransferases-Role and Function. Advances in Experimental Medicine and Biology. Dordrecht Heidelberg London New York: Springer Science+Business Media; 2016. pp. 61-84

[35] Adachi N, Lieber MR. Bidirectional gene organization: A common architectural feature of the human genome. Cell. 2002;**109**(7):807-809

[36] Yang MQ, Elnitski L. Diversity of core promoter elements comprising human bidirectional promoters. BMC Genomics. 2008;**9**(Suppl. 2):S3

[37] Trinklein ND, Aldred SF, Hartman SJ, Schroeder DI, Otillar RP, et al. An abundance of bidirectional promoters in the human genome. Genome Research. 2004;**14**(1):62-66

[38] Luo C, Hajkova P, Ecker JR. Dynamic DNA methylation: In the right place at the right time. Science. 2018;**361**(6409):1336-1340

[39] Docherty LE, Rezwan FI, Poole RL, Jagoe H, Lake H, et al. Genome-wide DNA methylation analysis of patients with imprinting disorders identifies differentially methylated regions associated with novel candidate imprinted genes. Journal of Medical Genetics. 2014;**51**(4):229-238

[40] Yin Y, Morgunova E, Jolma A, Kaasinen E, Sahu B, et al. Impact of cytosine methylation on DNA binding specificities of human transcription factors. Science. 2017;**356**(6337):eaaj2239

[41] Suzuki M, Yamada T, Kihara-Negishi F, Sakurai T, Hara E, et al. Site-specific DNA methylation by a complex of PU.1 and Dnmt3a/b. Oncogene. 2006;**15**(17):2477-2488

[42] Hervouet E, Vallette FM, Cartron PF. Dnmt1/transcription factor interactions: An alternative mechanism of DNA methylation inheritance. Genes & Cancer. 2010;**1**(5):434-443

[43] Sardina JL, Collombet S, Tian TV, Gómez A, Di Stefano B, et al. Transcription factors drive Tet2-mediated enhancer demethylation to reprogram cell fate. Cell Stem Cell. 2018;**23**(5):727-741

[44] Haro D, Marrero PF, Relat J. Nutritional regulation of gene expression: Carbohydrate-, Fat-, and acmino acid-dependent modulation of transcriptional activity. International Journal of Molecular Sciences. 2019;**20**(6):1386

[45] Mitro N, Mak PA, Vargas L, Godio C, Hampton E, et al. The nuclear receptor LXR is a glucose sensor. Nature. 2007;**445**(7124):219-223

[46] Tu AY, Albers JJ. Glucose regulates the transcription of human genes relevant to HDL metabolism: Responsive elements for peroxisome proliferator-activated receptor are involved in the regulation of phospholipid transfer protein. Diabetes. 2001;**50**(8): 1851-1856

[47] Zhou B, Ikejima T, Watanabe T, Iwakoshi K, Idei Y, et al. The effect of 2-deoxy-D-glucose on Werner syndrome RecQ helicase gene. FEBS Letters. 2009;**583**(8):1331-1336

[48] Latruffe N, Cherkaoui Malki M, Nicolas-Frances V, Clemencet MC, et al. Regulation of the peroxisomal β-oxidation-dependent pathway by peroxisome proliferator-activated receptor α and kinases. Biochemical Pharmacology. 2000;**60**(8):1027-1032

[49] Magaña MM, Osborne TF. Two tandem binding sites for sterol regulatory element binding proteins are required for sterol regulation of fatty-acid synthase promoter. The Journal of Biological Chemistry. 1996;**271**(51):32689-32694

[50] Jump DB, Tripathy S, Depner CM. Fatty acid-regulated transcription factors in the liver. Annual Review of Nutrition. 2013;**33**:249-269

[51] Vanhoutvin SA, Troost FJ, Hamer HM, Lindsey PJ, Koek GH, et al. Butyrate-induced transcriptional changes in human colonic mucosa. PLoS One. 2009;**4**(8):e6759

[52] Chriett S, Dąbek A, Wojtala M, Vidal H, Balcerczyk A, et al. Prominent action of butyrate over beta-hydroxybutyrate as histone deacetylase inhibitor, transcriptional modulator and anti-inflammatory molecule. Scientific Reports. 2019;**9**(1):742

[53] Torun A, Enayat S, Sheraj I, Tunçer S, Ülgen DH, et al. Butyrate mediated regulation of RNA binding proteins in the post-transcriptional regulation of inflammatory gene expression. Cellular Signalling. 2019;**64**:109410

[54] Wierstra I. Sp1: Emerging roles - beyond constitutive activation of TATA-less housekeeping genes. Biochemical and Biophysical Research Communications. 2008;**372**(1):1-13

[55] Waby JS, Chirakkal H, Yu C, Griffiths GJ, Benson RS, et al. Sp1 acetylation is associated with loss of DNA binding at promoters associated with cell cycle arrest and cell death in a colon cell line. Molecular Cancer. 2010;**15**(9):275

[56] Ginsburg E, Salomon D, Sreevalsan T, Freese E. Growth inhibition and morphological change caused by lipophilic acids in mammalian cells. Proceedings of the National Academy of Sciences of the United States of America. 1973;**70**(8):2457-2461

[57] Goel A, Janknecht R. Acetylation-mediated transcriptional activation of the ETS protein ER81 by p300, P/CAF, and HER2/Neu. Molecular and Cellular Biology. 2003;**23**(17):6243-6254

[58] Guo B, Panagiotaki N, Warwood S, Sharrocks AD. Dynamic modification of the ETS transcription factor PEA3 by sumoylation and p300-mediated acetylation. Nucleic Acids Research. 2011;**39**(15):6403-6413

[59] Gaub P, Tedeschi A, Puttagunta R, Nguyen T, Schmandke A, et al. HDAC inhibition promotes neural outgrowth and counteracts growth cone collapse through CBP/p300 and P/CAF-dependent p53 acetylation. Cell Death and Differentiation. 2010;**17**(9):1392-1408

[60] Brasse-Lagnel C, Lavoinne A, Husson A. Control of mammalian gene expression by amino acids, especially glutamine. The FEBS Journal. 2009;**276**(7):1826-1844

[61] Bruhat A, Jousse C, Carraro V, Reimold AM, Ferrara M, et al. Amino

acids control mammalian gene transcription: Activating transcription factor 2 is essential for the amino acid responsiveness of the *CHOP* promoter. Molecular and Cellular Biology. 2000;**20**(19):7192-7240

[62] Bender DA. Micronutrients. In: Bender DA, editor. Introduction to Nutrition and Metabolism. 5th ed. Boca Raton, FL: CRC Press, Taylor & Francis Group, Inc.; 2014. pp. 307-384

[63] Uchiumi F, Sato A, Asai M, Tanuma S. An $NAD^+$ dependent/sensitive transcription system: Toward a novel anti-cancer therapy. AIMS Molecular Science. 2020;**7**(1):12-28

[64] Ghosh S, George S, Roy U, Ramachandran D, Kolthur-Seetharam U. NAD: A master regulator of transcription. Biochimica et Biophysica Acta. 2010;**1799**(10-12):681-693

[65] Ryu KW, Nandu T, Kim J, Challa S, DeBerardinis RJ, et al. Metabolic regulation of transcription through compartmentalized $NAD^+$ biosynthesis. Science. 2018;**360**(6389):eaan5780

[66] Goodman RP, Markhard AL, Shah H, Sharma R, Skinner OS, et al. Hepatic NADH reductive stress underlies common variation in metabolic traits. Nature. 2020;**583**(7814):122-126

[67] Gall WE, Beebe K, Lawton KA, Adam KP, Mitchell MW, et al. α-Hydroxybutyrate is an early biomarker of insulin resistance and glucose intolerance in a nondiabetic population. PLoS One. 2010;**5**(5):e10883

[68] Rehmani I, Liu F, Liu A. Cell signaling and transcription. In: Villamena FA, editor. Molecular Basis of Oxidative Stress: Chemistry, Mechanisms, and Disease Pathogenesis. Hoboken, NJ: John Wiley & Sons; 2013. pp. 179-201

[69] Takihara Y, Sudo D, Arakawa J, Takahashi M, Sato A, et al. Nicotinamide adenine dinucleotide ($NAD^+$) and cell aging. In: Strakoš R, Lorens B, editors. New Research on Cell Aging and Death. Hauppauge, NY: Nova Science Publishers, Inc.; 2018. pp. 131-158

[70] Uchiumi F, Watanabe T, Hasegawa S, Hoshi T, Higami Y, et al. The effect of Resveratrol on the Werner Syndrome RecQ helicase gene and telomerase activity. Current Aging Science. 2011;**4**:1-7

[71] Liu J, Ali M, Zhou Q. Establishment and evolution of heterochromatin. Annals of the New York Academy of Sciences. 2020;**1476**(1):59-77

[72] van Steensel B, Belmont AS. Lamina-associated domains: Link with chromosome architecture, heterochromatin, and gene repression. Cell. 2017;**169**(5):780-791

[73] Dillon N. Heterochromatin structure and function. Biology of the Cell. 2004;**96**(8):631-637

[74] Nava MM, Miroshnikova YA, Biggs LC, Whitefield DB, Metge F, et al. Heterochromatin-driven nuclear softening protects the genome against mechanical stress-induced damage. Cell. 2020;**181**(4):800-817

[75] Fortuny A, Polo SE. The response to DNA damage in heterochromatin domains. Chromosoma. 2018;**127**(3):291-300

[76] Dantzer F, Santoro R. The expanding role of PARPs in the establishment and maintenance of heterochromatin. The FEBS Journal. 2013;**280**(15):3508-3518

[77] Smith S, Giriat I, Schmitt A, de Lange T. Tankyrase, a poly(ADP-ribose)

polymerase at human telomeres. Science. 1998;**282**(5393):1484-1487

[78] Mathieu N, Pirzio L, Freulet-Marrière MA, Desmaze C, Sabatier L. Telomeres and chromosomal instability. Cellular and Molecular Life Sciences. 2004;**61**(6):641-656

[79] Earle E, Saxena A, MacDonald A, Hudson DF, Shaffer LG, et al. Poly (ADP-ribose) polymerase at active centromeres and neocentromeres at metaphase. Human Molecular Genetics. 2000;**9**(2):187-194

[80] Naegeli H, Althaus FR. Regulation of poly(ADP-ribose) polymerase. Histone-specific adaptations of reaction products. The Journal of Biological Chemistry. 1991;**266**(16):10596-10601

[81] Kim HL, Ra H, Kim KR, Lee JM, Im H, et al. Poly(ADP-ribosyl)ation of p53 contributes to TPEN-induced neuronal apoptosis. Molecules and Cells. 2015;**38**(4):312-317

[82] Ding L, Chen X, Xu X, Qian Y, Liang G, et al. PARP1 suppresses the transcription of PD-L1 by poly(ADP-ribosyl)ating STAT3. Cancer Immunology Research. 2019;**7**(1):136-149

[83] Dykes IM, Emanueli C. Transcriptional and post-transcriptional gene regulation by long non-coding RNA. Genomics, Proteomics & Bioinformatics. 2017;**15**(3):177-186

[84] Bishop JM. Enemies within: The genesis of retrovirus oncogenes. Cell. 1981;**23**(1):5-6

[85] Elbarbary RA, Lucas BA, Maquat LE. Retrotransposons as regulators of gene expression. Science. 2016;**351**(6274):aac7247

[86] Ling X, Han Y, Meng J, Zhong B, Chen J, et al. Small extrachromosomal circular DNA (eccDNA): Major functions in evolution and cancer. Molecular Cancer. 2021;**20**(1):113

[87] Paulsen T, Kumar P, Koseoglu MM, Dutta A. New discoveries of extrachromosomal circles of DNA in normal and tumor cells. Trends in Genetics. 2018;**34**(4):270-278

[88] Peters JM. How DNA loop extrusion mediated by cohesin enables V(D)J recombination. Current Opinion in Cell Biology. 2021;**71**:75-83

Chapter 2

# Transcription Elongation Factors in Health and Disease

*Preeti Dabas*


## Abstract

Gene expression is a complex process that establishes and maintains a specific cell state. Transcription, an early event during the gene expression, is fine-tuned by a concerted action of a plethora of transcription factors temporally and spatially in response to various stimuli. Most of the earlier research has focused on the initiation of transcription as a key regulatory step. However, work done over the last two decades has highlighted the importance of regulation of transcription elongation by RNA Pol II in the implementation of gene expression programs during development. Moreover, accumulating evidence has suggested that dysregulation of transcription elongation due to dysfunction of transcription factors can result in developmental abnormalities and a broad range of diseases, including cancers. In this chapter, we review recent advances in our understanding of the dynamics of transcription regulation during the elongation stage, the significance of transcriptional regulatory complexes, and their relevance in the development of potential accurate therapeutic targets for different human diseases.


**Keywords:** transcription, RNA Pol II, elongation factor

## 1. Introduction

There are ~20,000 protein-coding genes in the human genome [1]. Cells modulate the expression of these genes in spatial and temporal manner in response to various stimuli. Gene expression is a highly regulated and complex process, which begins with the opening of chromatin, the transcription of the primary RNA transcript (hnRNA) from DNA, followed by processing of the hnRNA into mRNA, which is then translated into a protein that dictates cell functions. There has been an extensive study on precise control of gene expression at different stages by a plethora of factors leading to the concept of "gene-class specific" or selective gene control [2–6]. Tight control of gene expression is indispensable for normal cellular functions, and any dysregulation may lead to a wide range of diseases. The recent surge in knowledge and understanding of diseases that are caused by a mutation in regulatory sequences, transcription factors, cofactors, chromatin regulators, and non-coding RNA, such as diabetes, autoimmune disorders, neurological disorders, obesity, cardiovascular disease, and cancer, has altered our view of the underlying cause and primary focus of therapeutic targets.

It is imperative to understand the process of regulated gene expression to get insights into the mechanisms involved in the dysregulation of gene expression in various human diseases and to develop potential therapeutic targets for these diseases. Transcription has been considered the most important rate-limiting step during the expression of a gene. For a long time, initiation of transcription was considered as the key regulatory event during transcription and thus was the focus of major research in this field. However, for over a decade, the focus of research has shifted toward other steps during transcription, such as elongation and termination. Moreover, the regulation of transcription elongation has emerged to be the center of most therapeutic studies targeting fine-tuning of gene expression in diseases such as cancers.

In this chapter, I begin with a brief review of steps involved in gene regulation and the fundamentals of transcriptional control of gene expression. I further focus on a detailed understanding of regulation of transcription elongation by different transcription elongation factors and complexes and how a mutation or dysfunction of any of these factors contributes to altered transcription leading to progression of various diseases and cancer. I also highlight the recent advances in the development of precision tailored therapeutics by mediating transcriptional control of a gene.

## 2. Stages of regulation of gene expression

In eukaryotes, modulation of gene expression occurs at seven different steps (**Figure 1**).

### 2.1 Chromatin structure

DNA wraps around proteins called histones to form nucleosomes. Each nucleosome is further condensed into chromatin. The condensation of eukaryotic DNA in chromatin acts to suppress the expression of genes by acting as a physical barrier

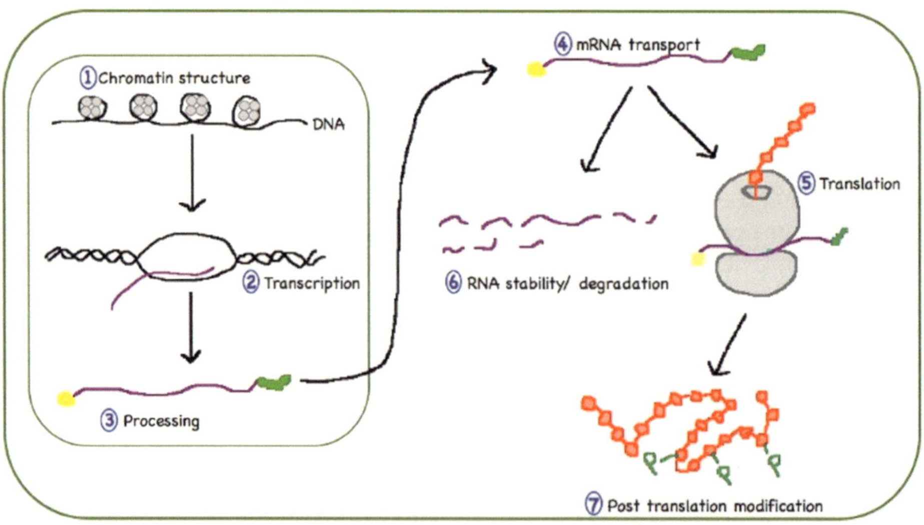


**Figure 1.**
*Schematic representation of steps involved in gene regulation. Regulation of gene expression occurs at seven different steps.*

to the transcription machinery [2]. The opening of chromatin, which allows access to genomic DNA, is indispensable for gene expression and formation of RNA from DNA template. This unwrapping of DNA from histone proteins is called chromatin remodeling and is carried out by enzymes that interact with histones and covalently attach functional groups to the amino terminal tail of histones. These histone-modifying proteins form complexes known as chromatin remodeling complexes. The most common modifications include methylation or acetylation of lysine residues on histone tail [3]. The outcome of these two modifications is entirely different: acetylation results in an open conformation of chromatin, thereby causing activation of gene expression; methylation results in a more compact chromatin conformation, hindering DNA accessibility to transcription factors and thus repressing transcription. Apart from histone modifications, methylation of DNA may also lead to a transcriptionally inactive state. A balance between the active and inactive state of chromatin is of profound importance for the maintenance of a healthy cellular environment. Dysfunctional chromatin remodeling complexes have been implicated in several disease conditions, including Williams syndrome [4, 5], Rett syndrome [6] breast cancer [7], and several other primary tumors [8].

### 2.2 Transcription

Transcription is the key regulatory step for gene expression in eukaryotes. It involves a concerted action of different proteins, such as transcription factors, mediator complex components, and RNA Pol II to produce RNA using DNA as a template. In eukaryotes, RNA Pol II is responsible for the synthesis of protein-coding genes as well as some non-coding RNAs such as small nuclear RNA (snRNA), microRNA (miRNA), cryptic unstable transcripts (CUTs), small nucleolar RNA (SnoRNA), and stable unannotated transcripts (SUTs) [9]. RNA Pol II mediated transcription is composed of three main steps: initiation, elongation, and termination, all of which are subjected to regulatory controls. During the stage of transcription initiation, the RNA Pol II in association with different transcription factors is recruited to the promoter region of the gene and forms a complex called pre-initiation complex (PIC). This opens the DNA, and the template strand positions itself in the active site of RNA Pol II, which then initiates the synthesis of the first few nucleotides of RNA. When the RNA length reaches ~10 nucleotides, RNA Pol II escapes the promoter and enters the gene body leading to productive elongation. Once Pol II reaches the end of the gene, RNA Pol II ceases RNA synthesis, which signals the release of the nascent RNA transcript as well as Pol II from the DNA template, thus terminating the process of transcription. All three stages of transcription are subject to tight control. Perhaps, due to its foremost position in the transcription cycle, the initiation step is extensively studied for mechanism and regulation. More recently, the transition of initiation to elongation has emerged to be the hub of major research in the field of transcription regulation. However, the mechanism and regulation of transcription termination have been less investigated.

A detailed description of regulation during transcription is described in the subsequent sections of this chapter.

### 2.3 RNA processing

The primary or nascent transcript is further processed to form functional mRNA. During processing, the primary transcript undergoes three types of modifications: 5′

end modification (capping), removal of introns (splicing), and 3′ end modification (polyadenylation) [10].

The newly synthesized RNA is stabilized by the addition of a 7-methylguanosine cap, which protects nascent from attack by nuclear exonucleases and helps in promoting transcription, splicing, polyadenylation, and nuclear export [11, 12]. Factors responsible for the capping of 5′ end of RNA, for example, eIF4E-antisense oligonucleotides, are being extensively used therapeutically in clinical trials that aim to curb dysregulated gene expression in cancer [13]. Small ribonucleoprotein particles (snRNPs) along with auxiliary proteins form a spliceosome complex, which mainly carries out the process of splicing by recognizing the splice sites and catalyzing the splice reaction [14]. Any dysregulation of the splicing mechanism results in diseased conditions [15]. Polyadenylation of 3'end of nascent RNA includes cleavage at polyadenylation site (PAS) of RNA and addition of poly (A) tail [16–18]. Alternate polyadenylation (APA) is yet another mechanism adopted by the cell to produce diversity in the mRNA pool. APA results in different isoforms of the same gene with varying 3'UTR [19]. Poly(A) tails are responsible for stability, translation efficiency, and degradation of RNA. Alteration in polyadenylation is associated with a plethora of diseased conditions, such as neonatal diabetes, fragile X-associated premature ovarian insufficiency, IPEX (immune dysfunction, polyendocrinopathy, enteropathy, X-linked), ectopic Cushing syndrome, and several cancer conditions such as endocrine tumor [20].

### 2.4 RNA transport

Once the mature RNA is made post RNA processing events, it is rearranged in an export competent mRNP complex with RNA-binding factors and shuttling proteins and transported to the cytoplasm. This process is tightly regulated. After the transport, in the cytoplasmic side, the DEAD-box helicase remodels the mRNP to dissociate RNA binding and shuttling proteins, preventing the mRNA from returning to the nucleus [21]. Furthermore, cap-binding protein (CBP), which binds to the 7-methylguanylate cap on 5′ end of processed RNA, is recognized by nuclear pore complex and exported to the cytoplasm, where it is replaced by translation factors eIF4E and eIF4G [21]. Few lines of research have shown that the transport machinery co-transcriptionally associates with mRNA [21–23], while others have shown that 3′ end processing of transcript marks the event responsible for the loading of export complex [24, 25]. For example, shuttling proteins, Hrp1p and Nab2p, associate with poly (A) tail [26, 27].

Any defect in mRNA transport or nuclear retention results in human disease such as lethal congenital contracture syndrome 1.

### 2.5 RNA stability and degradation

The life span of mRNA in cytosol determines the protein turnover. This is an important step to control gene expression since modulation of mRNA abundance allows cell to adapt and respond to various situations adequately. Generally, proteins with housekeeping functions are encoded by mRNA with a long half-life, while the proteins required only at specific developmental stages are encoded by mRNA with a short half-life [28, 29]. mRNA decay is a highly regulated process, resulting from interactions between mRNA, RNA-binding proteins, non-coding RNA, and various decay factors. The stringency of mRNA degradation depends on the presence of regulatory RNA elements, consisting of specific sequences found anywhere in mRNA, including 5′ and 3' UTR. These sequences are recognized by RNA-binding proteins, forming mRNP complexes [30]. These RNA-binding proteins determine the fate of bound mRNA—to be translated, decayed, or stored in untranslated form as cytoplasmic granules. mRNA degradation begins with deadenylation or shortening of poly(A) tail carried out by Pan2/Pan3 complex and Ccr4/Pop2/Not complexes. After deadenylation, the mRNA is degraded either by 3′ → 5' mRNA decay pathway mediated by exosome or 5′ → 3' mRNA decay pathway mediated by exonuclease Xrn1 after decapping by Dcp1/Dcp2 decapping complex [31]. More recent studies have focused on the role of non-coding RNA molecules called miRNA in regulation of mRNA degradation and subsequently gene expression. miRNA works by either repressing translation or promoting degradation of mRNA having sequence complementary to miRNA [32]. Dysregulation of mRNA stability has been implicated in several diseases, including tumors. For instance, in myeloma and human T-cell leukemia, the stability of c-Myc RNA (an oncogene) due to loss of 3'UTR, which is responsible for its decay, results in its stability up to 4–8-fold higher compared with the wild type [33, 34].

### 2.6 Translation

Regulation of translation is a crucial mechanism for spatial control of gene expression [35]. There has been an explosion of studies highlighting the importance of regulation of translation in various physiological processes such as in normal development [36], in apoptosis [37], in stress response [38], etc. Dysregulation of translation has been implicated in different cancers [39]. In cells lacking nucleus, such as neurons, regulation of gene expression by translation has been shown to be of utmost importance. Several studies have demonstrated the initiation of translation as one of the crucial regulatory steps of protein synthesis [38, 40–42]. The role of P-bodies and small RNAs in translation regulation has also been elucidated. Initially, P-bodies were discovered as small foci rich in mRNA decay enzymes [43–48]. When encountered with stress, yeast cells block translation initiation, as is evident by reduced polysome number and increased size of P-bodies [49]. Brengues et al. [50] have demonstrated that after the removal of stress and absence of new transcription, there is a decrease in the size of P-bodies and the reformation of polysomes [50]. This shows that P-bodies also act as storage sites for mRNA without undergoing degradation. On the other hand, small RNAs also the regulate translation and stability of mRNA [51, 52].

### 2.7 Posttranslational modifications and protein degradation

After its synthesis, the polypeptide is folded and modified by the addition of various chemical groups or by removal of certain amino acids from polypeptide (proteolytic cleavage). These protein modification steps act as targets for regulation. For example, phosphorylation of eIF2 results in inactivity and thus blocking of translation [53]. However, in some instances, phosphorylation may enhance the activity of a protein. Moreover, when not required by the cell, these proteins undergo degradation *via* the process of polyubiquitination, which is again regulated.

## 3. Regulation of transcription

In eukaryotes, there are three types of RNA polymerases responsible for transcription: RNA polymerase I is responsible for the synthesis of rRNA; RNA Pol II is responsible for the production of protein-coding mRNA, long non-coding RNA, snRNA, and miRNA; and RNA Pol III (RNA Pol III) is responsible for the synthesis of tRNA, some small non-coding RNA, 5S, and 5.8S RNA. Although transcription by all these enzymes is amenable to regulation, we will focus on the transcription of protein-coding genes in this chapter. Transcription is one of the most critical steps during gene expression and is regulated by a plethora of transcription factors working in a concerted manner with RNA Pol II to ensure proper initiation, elongation, and termination of transcription. RNA Pol II transcription involved initiation, elongation, and termination of transcription. For a long time, regulation of transcription was majorly focused on the initiation step. However, for over a decade, there has been an increase in mechanistic insights of regulation of transcription elongation, marking it as another regulatory event during transcription.

### 3.1 Transcription initiation

Initiation of transcription begins with recognizing and binding of RNA Pol II to the promoter. However, since RNA Pol II cannot recognize the promoter, it needs assistance from other proteins called as "general transcription factors" (GTFs) [54]. The GTFs are evolutionarily conserved proteins, and their ordered recruitment to RNA Pol II is necessary to initiate RNA synthesis. There are six types of GTFs: TFIIA, TFIIB, TFIID, TFIIE, TFIIF, and TFIIH. **Table 1** summarizes the function of these GTFs.

Recruitment of RNA Pol II and general transcription factors (known as PIC or "Pre-Initiation Complex") to the promoter is highly regulated during the initiation of transcription. **Figure 2** shows the process of transcription initiation in eukaryotes. Specific transcription factors bind to the regulatory regions of the promoter. They work by modulating the assembly and activity of transcription machinery, either through direct interaction with components of basal transcription machinery or through action on chromatin [55, 56]. The mediator complex facilitates the

| General Transcription Factor | Subunits | Function |
|---|---|---|
| TFIIA | 3 ($\alpha$, $\beta$, $\gamma$) | Interact with TBP subunit of TFIID and stabilize PIC |
| TFIIB | 1 | Help in transcription start site selection, recruitment of TFIIH/RNA polymerase IIcomplex and assist in promoter escape |
| TFIID | 15 [TBP (TATA box binding protein) and 14 TAFs (TBP associating factors)] | Recognize promoter by binding to TATA box, mediate interaction between activators and basal transcription machinery |
| TFIIE | 2 (TFA1, TFA2 in *Saccharomyces cerevisiae* or $\alpha$, $\beta$ in human) | Facilitate TFIIH recruitment |
| TFIIF | 3 in *S. Cerevisiae* (Tfg1, Tfg2, Tfg3); 2 in human (RAP74, RAP20) | Promote binding of RNA polymerase IIto TFIID-TFIIB DNA ternary complex, help in transcription start site selection |
| TFIIH | 10 [7 core subunits, 3 kinase modules (CAK)] | DNA dependent ATPase, ATP dependent DNA helicase and CTD kinase, involved in promoter escape, promoter proximal pausing, elongation, RNA processing and termination |

**Table 1.**
*General transcription factors in eukaryotes and their functions.*

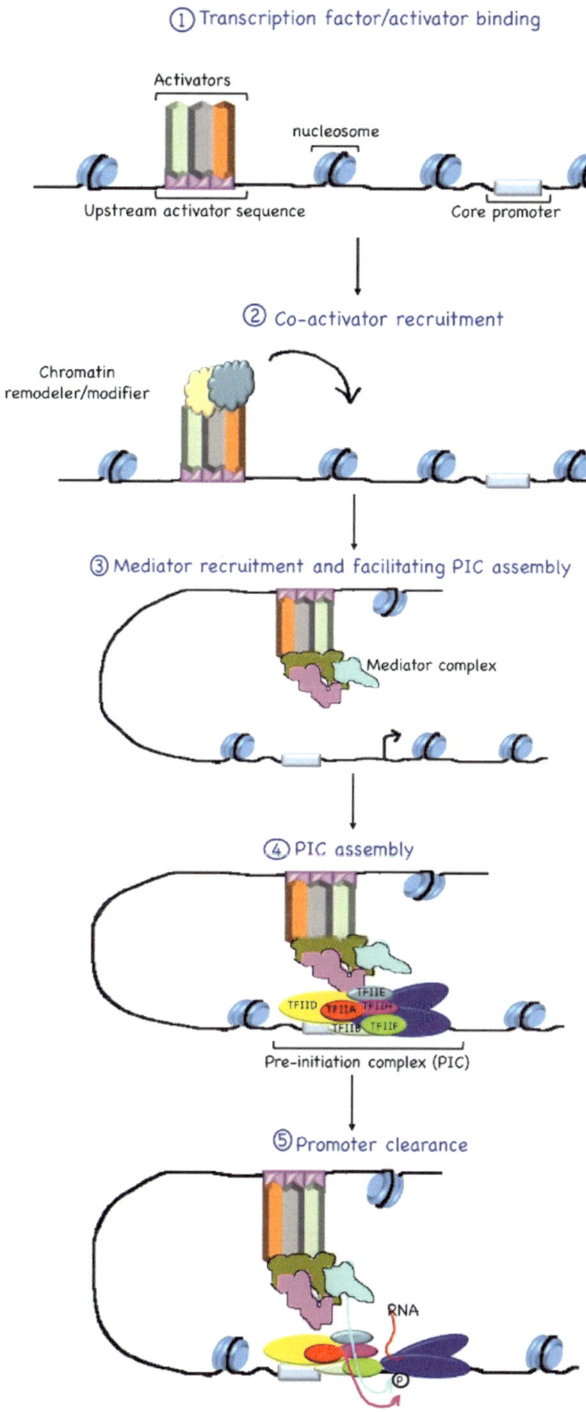


**Figure 2.**
*Transcription initiation. Diagram representing stepwise recruitment of factors leading to initiation of transcription. Transcription activators bind to the UAS, which recruits co-activators including chromatin remodeling complexes. This opens the chromatin and facilitates association of RNA Pol II along with GTFs at the gene promoter, forming PIC. Association between activators and PIC is mediated by mediator complex, which phosphorylates RNA Pol II at CTD and initiates the process of transcription.*

connection between the activators associated with UAS and the PIC bound to the promoter region. The role played by the mediator complex, activators, and repressors marks yet another event of regulation during the initiation of transcription. The mediator complex is a multiprotein complex primarily comprised of head, middle, tail, and kinase modules. The head and middle domain form the core of the mediator complex, while the tail and kinase domains serve as regulatory modules [57–60]. Although the mediator complex is conserved across evolution, the number of subunits vary in different species, comprising 19, 25, or up to 30 subunits in *S. pombe*, *Saccharomyces cerevisiae,* or humans, respectively. Five subunits have been reported to be metazoan-specific: MED23, MED25, MED26, MED28, and MED30 [61]. **Figure 3** represents the subunit composition of the mediator complex in yeast and human. The mediator complex functions as a bridge between basal transcription machinery and specific transcription factors, resulting in the assembly of the pre-initiation complex (PIC) at the core promoter [62]. Besides recruitment of basal transcription machinery during initiation, the role of the mediator complex has been established in transition from initiation to elongation [63], during elongation [64, 65], as well as during termination [66], mRNA export [67], DNA repair [68], and *S. pombe* cell separation [69].

### 3.2 Transcription elongation

Promoter clearance is the transit phase between transcription initiation and elongation. Several factors determine the ability of RNA Pol II to move out of the promoter and enter the elongation phase. Co-crystallization of RNA Pol II with TFIIB has demonstrated that TFIIB obstructs the exit channel for newly synthesized RNA, and therefore, removal of TFIIB is imperative to promoter escape [70]. Transcription elongation is divided into two phases: early elongation and productive elongation [71]. The phosphorylation status of CTD of RNA Pol II is also an important determinant of the stage of transcription: during PIC assembly and initiation of transcription, CTD remains unphosphorylated [72, 73], whereas phosphorylation at serine 5 marks promoter clearance [74]. Serine 5 phosphorylated RNA Pol II associates with promoter-proximal regions during transcription initiation and

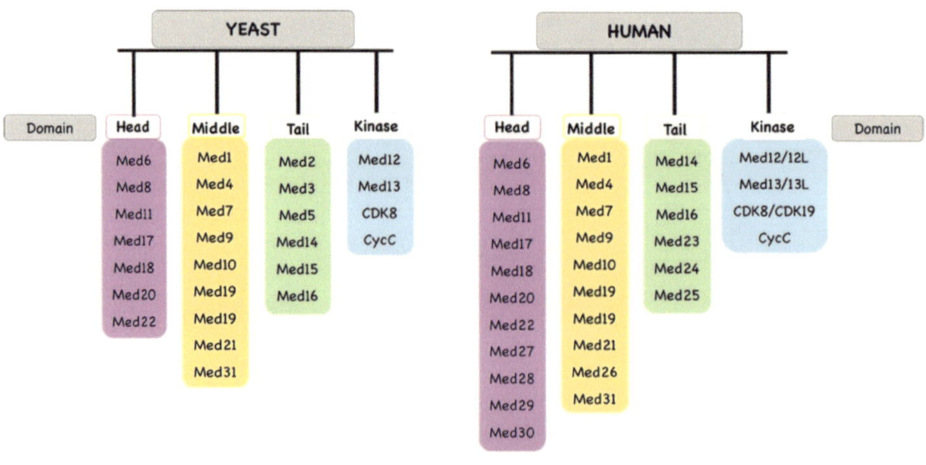


**Figure 3.**
*Subunit composition of mediator complex in yeast and human. Diagram representing different modules comprising various subunits of mediator complex in yeast and human.*

early elongation, while serine 2 phosphorylation is associated with distal promoter regions during transcription elongation [75]. The CDK8 subunit of kinase module of mediator complex and CDK7 subunit of TFIIH are responsible for phosphorylation of CTD during initiation and early elongation phase, signaling RNA Pol II to clear promoter [76]. Phosphorylation of serine 2 on RNA Pol II CTD by P-TEFb triggers the passage into productive elongation from the early elongation phase [77, 78].

During the early elongation phase, RNA Pol II encounters various hurdles including transcriptional pause, arrest, or termination. The phenomenon of "promoter-proximal pausing" is characterized by transient pausing of RNA Pol II after synthesis of 20–60 nucleotides long RNA before resuming transcription elongation [71, 79]. Promoter-proximal pausing is well established in metazoans but less frequently observed in yeast [80, 81].

Evidence from biochemical studies has shown that pausing/arrest occurs as a result of backtracking of RNA Pol II on DNA template, thereby displacing the 3′ end of nascent RNA from the active site in RNA Pol II. This process can be spontaneously reversed ("pausing") or not ("arrest") [82]. The release of RNA Pol II from pause/arrest has emerged as an important mechanism to ensure continued and effective transcription elongation.

#### *3.2.1 Transcription elongation factors*

Biochemical experiments have shown that purified RNA Pol II proceeds optimally at rates of only 100–300 nucleotides/min in vitro as compared with the rate of 1200–2000 nucleotides/min in vivo [83–85]. The in vitro slow rate of mRNA synthesis was reported to be due to frequent pausing or arrest of RNA Pol II along the DNA [86, 87], suggesting elongation to be an inherently discontinuous process. An array of proteins known as "transcription elongation factors" function to regulate the rate of elongation.

These transcription elongation factors have been classified into different classes:

- Factors that assist RNA Pol II to traverse through transient pausing sites, e.g., P-TEFb [73], DRB sensitivity inducing factor [88, 89];
- Factors that can assist RNA Pol II to transcribe through chromatin, e.g., FACT, Swi/Snf [90, 91].
- Factors that can increase the overall rate of transcribing RNA Pol II, e.g., Elongin [92, 93], ELL [94, 95], ELL2 [96].
- Factors that suppress the activity of RNA Pol II, e.g., NELF [97].

Some transcription elongation factors increase transcription of all protein coding genes, while others expedite transcription of only a set/class of genes [86]. Moreover, there are several GTFs as well as elongation factors that regulate transcription during either early or productive elongation phase of transcription.

#### *3.2.2 Transcription factors regulating early elongation phase and pause release*

##### *3.2.2.1 TFIIF*

TFIIE, TFIIF, and TFIIH have been implicated in post-initiation functions regulating early elongation and promoter escape during transcription [98].

In mammals, it is composed of two subunits, RAP30 and RAP74 [99]. Documented evidence suggests that TFIIF functions in suppressing RNA Pol II-associated transient pausing during active elongation *via* direct interaction with the ternary complex [100, 101]. Several lines of research have established that distinct functions of TFIIF during initiation and elongation are carried out by its different functional domains [100–102]. For instance, the initiation activity of TFIIF is mediated by a DNA binding domain in the C-terminal region of RAP30 while the elongation activity is carried out by the RNA Pol II binding regions in the upstream sequence of the RAP30 C-terminus [98].

#### *3.2.2.2 TFIIH*

TFIIH is a conserved protein composed of 10 subunits, seven of which form the core, while three subunits comprise a catalytic module called CDK activating kinase (CAK) comprising CDK7, ATP-dependent helicase XBP, and XPD [103].

The regulated recruitment of TFIIH to the promoter is orchestrated by TFIIE [104–106]. TFIIE and TFIIH work together in suppressing premature arrest of the early RNA Pol II, thereby facilitating promoter escape [98, 107]. The CDK7 phosphorylates CTD and initiates elongation [108]. Furthermore, it has been shown that the CDK7 subunit of TFIIH recruits ELL at sites of DNA damage and helps in transcription restart after repair of damaged DNA [109].

#### *3.2.2.3 DSIF*

A nucleoside analog, 5,6-dichloro-1-β-D-ribofuranosylbenzimidazole (DRB), works by obstructing the transition from initiation to elongation by inhibiting the phosphorylation of CTD of RNA Pol II [110–112]. DRB sensitivity inducing factor (DSIF) was initially identified from HeLa cell nuclear extracts as a transcription factor that promotes pausing of RNA Pol II in response to DRB [88, 97]. The two subunits of DSIF, Spt4 and Spt5, regulate the activity of RNA Pol II by genetically and physically interacting with it [88, 89, 97, 113, 114]. Association of DSIF with RNA Pol II after promoter escape is controlled by CDK7 dependent phosphorylation of the Spt5 subunit [115, 116]. Reduction in levels of the nascent and processed snRNA transcripts upon knockdown of DSIF in HeLa cell lines has pointed toward the function of DSIF as a transcriptional elongation regulator [117].

#### *3.2.2.4 NELF*

Negative elongation factor (NELF) is composed of five subunits, namely NELF-A, NELF-B, NELF-C, NELF-D, and NELF-E [118]. Interestingly, NELF is essential in *Drosophila melanogaster* and mammalian cells but is absent in *Saccharomyces cerevisiae*, *C. elegans*, and *Arabidopsis thaliana* [119, 120]. NELF promotes pausing of RNA Pol II by binding to RNA Pol II-DSIF complex through its NELF-E subunit [97, 118]. Yamaguchi et al. [97] demonstrated that DSIF/NELF interacts with the hypophosphorylated CTD of RNA Pol II to suppress its elongation function [97]. DSIF/NELF-mediated pausing of RNA Pol II gives sufficient time for recruitment of capping enzyme and addition of 5'cap to the nascent RNA [119, 121].

#### *3.2.2.5 P-TEFb*

Positive transcription elongation factor (P-TEFb) was first identified from a partially purified transcription system as a kinase inhibited by DRB [97]. It is composed of two subunits: cyclin T and CDK9. P-TEFb alleviates the NELF/DSIF-induced RNA Pol II pausing by phosphorylation of Spt5 subunit of DSIF and Ser2 residue of CTD heptad resulting in dissociation of NELF and transition from paused state to productive elongation state [71, 97, 119, 122–124]. Moreover, the FACT complex cooperates with P-TEFb to mitigate NELF/DSIF-mediated inhibition of transcription elongation [125]. In the cell, P-TEFb is subjected to stringent regulation and exists in an active as well as the inactive state. In mammals, most of the inactive P-TEFb exists as a part of complex, which includes an inactive ribonucleoprotein complex called 7SK snRNP, which consists of 7SK snRNA, P-TEFb, HEXIM1/2 (hexamethylene bisacetamide inducible protein), LARP7 (La-related protein 7), and MePCE (methyl phosphate capping enzyme) [126, 127]. However, when rapid transcription induction is required, P-TEFb is released from the inactive complex and is recruited to the transcription site by a specific activator or chromatin remodeling protein, bromodomain-containing protein 4 (BRD4) [79, 128, 129]. P-TEFb is a component of yet another multiprotein complex called super elongation complex (SEC), which is involved in productive elongation by RNA Pol II [130]. P-TEFb has been implicated as a potential therapeutic target in multiple myeloma, autoimmune diseases, cardiac hypertrophy, and infectious diseases [131–137]. An emerging concept of indirect therapeutic targeting using synthetic transcription elongation factor has shown enhanced gene expression in diseased conditions where a particular gene is downregulated. A synthetic transcription elongation factor is composed of a programmable DNA-binding ligand attached to a small molecule that can bind to and recruit the transcription elongation machinery, thereby regulating the gene expression. Syn-TEF1 has been shown to engage P-TEFb via recruitment of BRD4 at GAA repeats and restored the expression of the FXN gene in Fredrich Ataxia cell line to the levels observed in healthy cells [138].

**Figure 4** depicts the role of P-TEFb in regulating the transition from the early elongation phase to productive elongation.

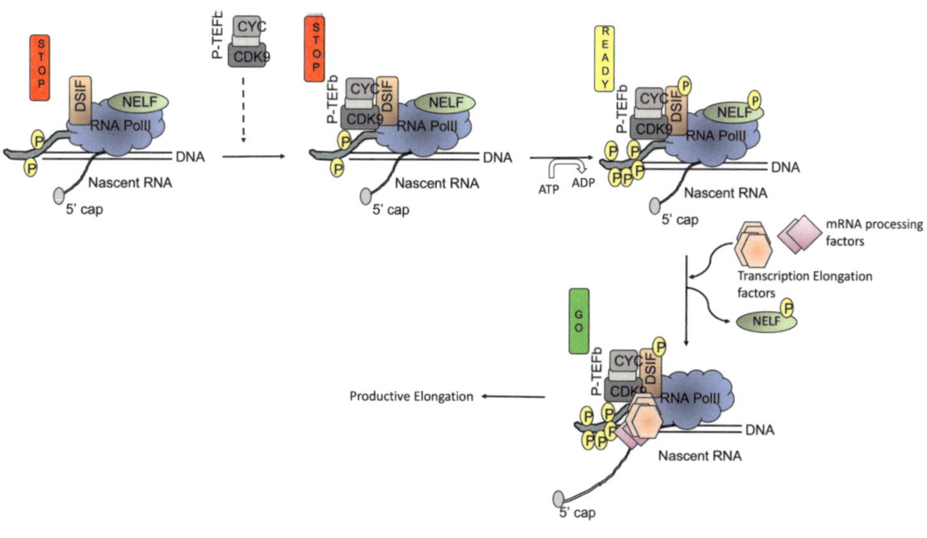


**Figure 4.**
*Role of P-TEFb in regulating transcription elongation. (a) P-TEFb is recruited to the paused RNA Pol II as a result of the negative effects of DSIF and NELF. (b) At pause sites, P-TEFb phosphorylates RNA Pol II CTD as well as NELF and DSIF. (c) Due to phosphorylation, NELF dissociates, DSIF functions as a positive factor in the absence of NELF, recruits other factors for mRNA elongation and processing, resulting in productive elongation.*

### *3.2.3 Transcription factors regulating productive elongation phase*

A separate class of transcription factors regulate the process of transcription during productive elongation, which is described below:

#### *3.2.3.1 Elongin*

Elongin increases the overall rate of transcription by suppressing the transient pausing of RNA Pol II through its interaction with RNA Pol II and stabilizing it in an active conformation for an extended period [139, 140]. Elongin is composed of three subunits: Elongin A, Elongin B, and Elongin C. Elongin A was found to be enriched at the transcriptionally active sites in association with an active form of RNA Pol II on polytene chromosomes of *Drosophila melanogaster* [141]. Elongin B and C were also found to be components of various Cullin2/Cullin5-based ubiquitin ligase complexes and interact with different proteins in the complex *via the* BC box motif. Elongin BC serves as an adaptor linking Elongin A to Cul2/Cul5 and RING finger protein Rbx1/2 containing modules [142]. Several studies have shown that the association of Elongin ABC with Cul5/Rbx2 contributes significantly to the degradation of stalled RNA Pol II at sites of DNA damage [143, 144].

#### *3.2.3.2 CSB*

A mutation in Cockayne syndrome A (CSA) or Cockayne syndrome B (CSB) genes results in an accelerated neurodegenerative disorder called cocaine syndrome [145, 146]. Clinical studies using cells from Cockayne syndrome patients exhibited a defect in transcription coupled repair (TCR) but not in global genome repair, indicating the role of CSA/CSB in TCR [147, 148]. CSB was shown to be transiently associated with DNA, RNA Pol II, and nascent RNA, and *in vitro* studies have shown that purified CSB stimulated the rate of elongation by RNA Pol II [139, 149]. Transcribing RNA Pol II stalls upon detecting lesions in DNA [150]. Blocked RNA Pol II is either retracted or dissociated by CSA and CSB proteins, thereby making the damage site on DNA accessible to repair proteins, accomplishing DNA damage repair and augmenting the resumption of transcription [151–153].

#### *3.2.3.3 TFIIS*

TFIIS, a zinc finger transcription factor, is known to stimulate the rate of RNA transcript synthesis. TFIIS is required for stimulating transcription elongation by reducing pause time, and it also increases the processivity of RNA Pol II on nucleosomes as well as stimulates translational elongation [124, 154, 155]. Interaction of TFIIS with ELL-Associated Factor 2 (EAF2) promotes transactivation by FESTA/EAF2 in murine embryonic stem cells [156].

#### *3.2.3.4 ELL family*

ELL (Eleven nineteen Lysine rich Leukemia) was first identified as a translocating partner to trithorax-like mixed-lineage leukemia (MLL) gene, located on 11q23 chromosomal locus observed in acute myeloid leukemia [157]. Functional characterization and mechanistic studies have shown that ELL plays a role during recruitment of PIC, promoter clearance, and release of RNA Pol II from pause sites, thereby stimulating the overall rate of transcription [94, 158]. The functions of ELL in connection with various disease conditions have also been reported. The expression of HIF-1α, as well as its downstream target genes, is elevated in the absence of ELL as observed in PC3 prostate cancer cell lines [159]. The Tax protein of Human T-cell Leukemia Virus Type 1 (HTLV-1) interacts with ELL and incorporates it into the p300 and P-TEFb containing complexes, which enhance the transcription of immediate early genes [160].

In humans, two other ELL homologs, namely ELL2 and ELL3, were identified based on sequence similarity with ELL [124]. ELL2 regulates pre-mRNA processing by assisting RNA Pol II to select weak promoter-proximal poly(A) sites in immunoglobulin heavy gene (IgH) [161–163]. ELL2 has been found to be upregulated in neuroendocrine prostate tumor, while its mRNA level was reduced in prostate adenocarcinoma and multiple myeloma plasma cells [164, 165]. The role of ELL3 has been implicated in breast cancer and B-cell lymphoma [166, 167].

#### *3.2.3.5 EAF family*

Yeast two-hybrid screens are carried out to identify proteins that associate with human ELL and identified two ELL interacting partners, namely EAF1 and EAF2 (ELL Associated Factor) [168, 169]. EAF2, also known as androgen-upregulated 19 (U19), was recognized as a novel testosterone-regulated protein that induces apoptosis in the prostrate [170]. EAF family of proteins act as cofactors to ELL, stimulating its transcription elongation activity in vitro [171]. The importance of EAF in diseases was first highlighted by a study that showed that the association of EAF with ELL is essential for the immortalization of hematopoietic progenitor cells and the development of acute myeloid leukemia [169, 172]. Since then, there has been an explosion in the studies describing the role of EAF1 and EAF2 in the development and progression of various tumors. A reduced expression of EAF2 in human prostate cancer specimen, lower survival rates upon complete loss of EAF2, and increased cell migration and proliferation in prostate tumor cell lines upon EAF2 knockdown underscore the role of EAF2/U19 as a tumor suppressor in the prostate [173–177]. A murine model for EAF2 knockout has further implicated the role of EAF2 in other diseased conditions such as B-cell lymphoma, hepatocellular carcinoma, prostate intraepithelial neoplasia (PIN), lung adenocarcinoma, and enlarged cardiac cells with an abnormal vascular system as well as abnormalities in spermatogenesis [178, 179]. EAF2 knockdown is also associated with heightened humoral immune response and excessive generation of autoantibody. The immune balance function of EAF2 is mediated by apoptosis of germinal center B cells [180]. A study has identified a frameshift mutation that resulted in the loss of EAF2 function in colorectal cancer as well as gastric cancer [181]. Recently, a study established that absence of EAF1 in mouse prostate triggers pre-neoplastic prostatic intraepithelial neoplasia lesions. However, a combined loss of both EAF1 and EAF2 significantly enhanced the proliferation and inflammation in the murine prostate and resulted in a more aggressive tumor when compared with the individual loss of either EAF1 or EAF2, indicating that coordination between the two homologs is required for maintaining homeostasis in the prostate [182].

### *3.2.4 Elongation complexes*

Several studies have demonstrated the existence of different multiprotein complexes called "Super Elongation Complex" (SEC), which are recruited to RNA Pol II to enhance its catalytic activity and thereby the rate of transcription elongation. These complexes are composed of transcription elongation factors such as p-TEFB, ELL1/2/3, EAF1/2 along with other proteins such as AFF4, ENL, AFF1, AFF9 [79]. In different organisms, super elongation complexes vary in composition and display functional diversity. For instance, in mammals, different combinations of four members of AFF family proteins (AFF1-4) and three members of ELL family protein (ELL, ELL2, and ELL3) confer functional variation to the SEC and form SEC-like complexes, SEC-L2 and SEC-L3 as shown in **Figure 4**. In place of AFF1/4, AFF2 and AFF3 represent the AFF family component of SEC-L2 and SEC-L3, respectively. Interestingly, biochemical studies have shown the absence of the one or more ELL family members in SEC-L2 and SEC-L3. The AFF4 and ELL2 containing SEC have been implicated in transcription elongation checkpoint control (TECC) in mammals. A similar super elongation complex was detected in *D. melanogaster*, comprising ELL, EAF1, P-TEFb, Lilli (AFF family member), and EAR (ENL) proteins [183]. Additionally, there is another complex present in *D. melanogaster*, known as Little Elongation Complex (LEC), comprising ICE1, ICE2, ELL, and EAF [184]. The role of LEC is found both in the initiation and elongation of snRNA transcription. Yet another elongation complex identified in hematopoietic cells called as "AEP complex" contains P-TEFb, AFF4, Afq31, and ENL protein [185]. The role of SEC in releasing paused RNA Pol II is very well described by several studies. However, other studies have also identified the role of SEC in rapid transcription by non-paused RNA Pol II in *Drosophila* embryos and in mouse embryonic stem cells [186, 187]. Recent work by Gopalan et al. [188] has recognized a rudimentary SEC in *S. pombe* consisting of only three members, ELL, EAF, and a newly identified AFF4 homolog EBP1 (ELL binding protein) [188]. The SECs were first reported two decades ago by their association with the viral transactivator of transcription (Tat) protein of HIV-1 and MLL-fusion partners involved in leukemogenesis [183, 189–192]. During HIV infection, the Tat protein interacts with the P-TEFb component of SEC and recruits it to the HIV-1 long terminal repeat (LTR) to stimulate the expression of HIV-1 in host cells. Different AFF family members dictate gene-class specific recruitment of SEC [193]. For instance, AFF1 containing SEC is important for interaction with Tat protein in HIV pathogenesis. On the other hand, AFF4 containing SEC is involved in Hsp70 gene expression upon heat shock. Since several MLL fusion partners are components of SEC, the fusion protein-mediated recruitment of SEC to hox genes points toward dysregulation of developmental genes in leukemia [183, 191, 192]. Furthermore, mutations in SEC components also result in dysregulation of the transcription elongation process leading to several diseases. A missense mutation in AFF4 results in CHOPS syndrome, a developmental disorder [194]. Given the importance of regulation of transcription elongation in gene expression in healthy and diseased states, several research groups have worked toward delineating the roles of transcription elongation complexes as therapeutic targets in diseases including cancers. Small-molecule inhibitors targeting SEC, such as KL-1 and KL-2, have been implicated in targeted therapy of myc-driven tumors [195].

### 3.3 Transcription termination

In mammals, the transcription termination of protein-coding genes is mainly dependent on termination complex or "CPSF-CF complex" comprising cleavage and polyadenylation specificity factor (CPSF or CPF in yeast), cleavage stimulating factor (CstF or CF1A in yeast), cleavage factor I (CFI), and cleavage factor II (CFII) [196, 197].

CPSF directly binds to the body of Pol II and recognizes polyadenylation signal (PAS) (AAUAAA). Association of CPSF with PAS and Pol II triggers transcription pausing. CstF, which associates with Ser2 phosphorylated residues on Pol II CTD, recognizes GU-rich processing signal downstream of PAS and facilitates cleavage of transcripts. This dislodges the CPSF, and RNA Pol II is released from pause. In eukaryotes, two models have been proposed to facilitate the termination of Pol II-mediated transcription after cleavage of nascent RNA [196–198]. The allosteric model postulates that the transcription terminates following destabilization of the elongation complex triggered by the loss of elongation factors/conformational change in Pol II after transcription of PAS. The second "Torpedo model" posits that the transcription termination occurs due to the dismantling of Pol II elongation complex following the degradation of nascent RNA by exonuclease. SETX (Sen1 in yeast) resolves the R-loop formed by short RNA left after cleavage and DNA. This allows the recruitment of 5′-3′ exoribonuclease XRN2 (Rat1 in yeast), which chews up the nascent RNA downstream of the cleavage site and releases Pol II from DNA [198, 199]. However, an emerging view in this field is that the combination of these two models likely explains the process of termination [200, 201]. The mechanism of transcription termination is depicted in **Figure 5**. A detailed mechanistic understanding of transcription termination is still missing, despite a surge in studies highlighting the role of transcription termination in controlling gene expression.

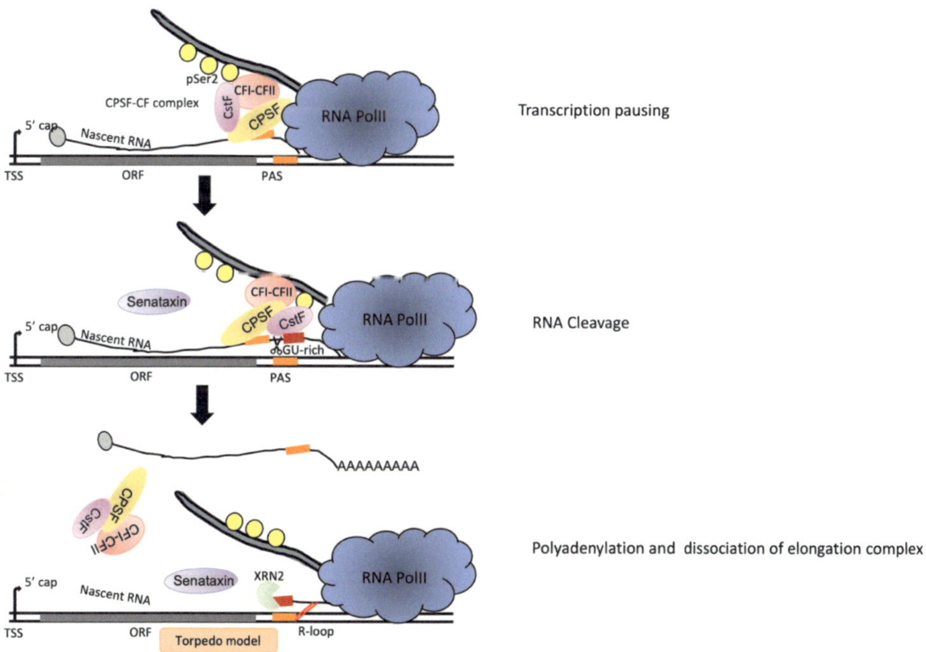


**Figure 5.**
*Mechanism of transcription termination at protein-coding genes in metazoans. Termination of transcription is triggered by the recruitment cleavage and polyadenylation factor complex (CPSF-CF complex) on the transcript. When RNA Pol II transcribes polyadenylation signal (PAS), the CPSF subunit binds to the PAS sequence on RNA while remaining bound to Pol II body. This results in pause of RNA Pol II. CstF subunit recognizes the GU-rich sequence downstream of PAS and creates a conformational change in the CPSF-CF complex, dissociating CPSF from Pol II and cleaving the nascent transcript between PAS and GU-rich sequence. R-loops formed by leftover transcript are resolved by Senataxin, allowing subsequent recruitment of XRN2 exonuclease. The remaining transcript downstream of the cleavage site is chewed up releasing Pol II and elongation complex by torpedo mechanism, thereby terminating the process of transcription.*

### 3.4 Recent advances in regulation of gene expression

Numerous studies have shown that transcription is carried out in membraneless phase-separated compartments when biomacromolecules such as proteins and transcription factors undergo self-assembly *via* liquid-liquid-like phase separation (LLPS). These phase-separated condensates act as a hub of transcription, where specific transcription factors are exchanged to ensure proper temporal and spatial gene expression patterns [202]. It has been reported that Pol II forms phase-separated condensate at active genes through its low-complexity CTD to concentrate transcription regulators and initiate the process of transcription [203–205]. SEC can also dynamically extract P-TEFb from inactive HEXIM1-P-TEFb complex through phase separation, resulting in activation of transcription elongation. This was confirmed by disruption of SEC phase-separated droplets, which reduces the transcription efficiency [206]. The roles of LLPS in disease states have also been explored recently. Oncogenic fusion of ENL, a component of SEC, and MLL has been shown to enhance the phase separation capabilities contributing to overactivation of leukemic genes [206]. Mutation in the YEATS domain of ENL has also been shown to promote self-assembly of ENL into LLPS, resulting in augmented recruitment of SEC to promote transcription of oncogenes in Wilms' tumor [207].

Another emerging concept of transcription regulation is through non-coding RNA and enhancer RNA (eRNA). Gene expression is regulated at multiple levels by long non-coding RNA (lncRNA): modulating chromatin structure and function, regulating transcription of neighboring or distal genes, RNA stability and translation [208–214]. The role of lncRNA has also recently been described in the regulation of nuclear condensates. Owing to the functional significance of lncRNA in transcription regulation, their role in the progression of diseases such as cancers and neurological disorders has been extensively studied. The transcription repression of tumor suppressors such as INK4A/ARF/INK4B has been associated with lncRNA ANRIL. ANRIL works by recruiting PRC1 and PRC2 complexes to promoters of these genes. Any dysregulation in ANRIL function may lead to silencing of these tumor suppressor genes contributing to tumor progression [215–218]. BACE1-AS, an antisense of gene encoding BACE1 protein, a precursor of amyloid plates in Alzheimer's disease (AD), promotes stability of BACE1 mRNA leading to accumulation of amyloid plates in the brains of AD patients. BACE1-AS also serves as a biomarker for AD and could be a potential therapeutic target to treat AD [219, 220]. lncRNAs are also involved in the suppression of gene expression through altered recruitment of transcription factors or Pol II or through reduced chromatin accessibility. For instance, following nerve injury, the lncRNA *Silc1* is necessary for activation of the SOX11 transcription program for nerve regeneration [221].

## 4. Conclusion

Regulation of gene expression is imperative to the normal physiological functioning of the cell. In recent years, we have observed remarkable progress in our understanding of the regulation of gene expression. Transcriptional regulation has emerged to be the most critical and well-studied stage during the expression of a gene. In line of its foremost position in the transcription process, the initiation step is most extensively researched and was considered a major rate-limiting step during transcription. However, now the focus has shifted from activation to elongation stage, which has been established as another key regulatory event during gene expression under normal physiological state. Regulation of transcription termination has only recently started to become the focus of several studies, and more mechanistic insights are required to fully understand regulatory events during this stage of transcription.

Accumulating evidence in the last few years has suggested the prevalence of "gene-class specific" transcription elongation factors, adding another layer to transcriptional regulation. These transcription elongation factors have been reported to be part of several multiprotein elongation complexes, which enhance the probability of cross talk between these factors and increase the regulatory potential of cells. Identification of phase-separated assemblies of transcription complexes has provided a biophysical basis of dynamic regulation of transcription in response to cellular cues. Another less-explored layer of transcription regulation is through non-coding RNA. Although there has been a tremendous increase in our understanding of the regulatory capabilities of lncRNA, we still lack a rigorous investigation to relate sequence and structural features of non-coding RNA to their regulatory functions.

A recent surge in targeted gene therapy has opened the doors for therapeutic targeting of "gene-class specific" transcription elongation factors in various diseases including cancers. An interesting concept in therapeutic targeting is of synthetic transcription elongation factors, which modulate the expression of a particular gene by selectively engaging the transcription elongation machinery at a specific gene locus. However, more efforts and research are required to dwell into the genome-wide perturbation of gene expression as a result of the binding of synthetic transcription factors to other less specific loci. In the context of personalized medicines, disease-related non-coding RNAs are gaining attention due to their specific expression patterns, which makes them a good candidate for disease biomarkers. Growing mechanistic insight into the regulation of transcription elongation and the interplay between different steps of regulation of gene expression would offer new aspects for intervention with aberrant modulation of gene expression and precisely tuned therapeutics.

## Acknowledgements

The author would like to thank Mr. Anurag Saroha for his help with the figures.

## Author details

Preeti Dabas
St. Jude Children's Research Hospital, Memphis, TN, USA

*Address all correspondence to: preeti.dabas3@gmail.com

## References

[1] Ezkurdia I, Juan D, Rodriguez JM, Frankish A, Diekhans M, Harrow J, et al. Multiple evidence strands suggest that there may be as few as 19 000 human protein-coding genes. Human Molecular Genetics. 2014;**23**(22):5866-5878

[2] Cosgrove MS, Boeke JD, Wolberger C. Regulated nucleosome mobility and the histone code. Nature Structural & Molecular Biology. 2004;**11**(11):1037

[3] Kouzarides T. Chromatin modifications and their function. Cell. 2007;**128**(4):693-705

[4] Hendrich B, Bickmore W. Human diseases with underlying defects in chromatin structure and modification. Human Molecular Genetics. 2001;**10**(20): 2233-2242

[5] Huang C, Sloan EA, Boerkoel CF. Chromatin remodeling and human disease. Current Opinion in Genetics & Development. 2003;**13**(3):246-252

[6] Zoghbi HY. Postnatal neuro-developmental disorders: Meeting at the synapse? Science. 2003;**302**(5646): 826-830

[7] Bochar DA, Wang L, Beniya H, Kinev A, Xue Y, Lane WS, et al. BRCA1 is associated with a human SWI/SNF-related complex: Linking chromatin remodeling to breast cancer. Cell. 2000;**102**(2):257-265

[8] Davis PK, Brachmann RK. Chromatin remodeling and cancer. Cancer Biology & Therapy. 2003;**2**(1):23-30

[9] Jacquier A. The complex eukaryotic transcriptome: Unexpected pervasive transcription and novel small RNAs. Nature Reviews Genetics. 2009;**10**(12): 833-844

[10] Bassett CL. Control of gene expression by mRNA transport and turnover. In: Regulation of Gene Expression in Plants. Boston, MA: Springer; 2007. pp. 148-188

[11] Lewis JD, Izaurflde E. The role of the cap structure in RNA processing and nuclear export. European Journal of Biochemistry. 1997;**247**(2):461-469

[12] Gu M, Lima CD. Processing the message: Structural insights into capping and decapping mRNA. Current Opinion in Structural Biology. 2005;**15**(1):99-106

[13] Graff JR, Konicek BW, Vincent TM, Lynch RL, Monteith D, Weir SN, et al. Therapeutic suppression of translation initiation factor eIF4E expression reduces tumor growth without toxicity. The Journal of Clinical Investigation. 2007;**117**(9):2638-2648

[14] Black DL. Mechanisms of alternative pre-messenger RNA splicing. Annual Review of Biochemistry. 2003;**72**(1): 291-336

[15] Chen M, Manley JL. Mechanisms of alternative splicing regulation: Insights from molecular and genomics approaches. Nature Reviews Molecular Cell Biology. 2009;**10**(11):741

[16] Shepard PJ, Choi EA, Lu J, Flanagan LA, Hertel KJ, Shi Y. Complex and dynamic landscape of RNA polyadenylation revealed by PAS-Seq. RNA. 2011;**17**(4):761-772

[17] Derti A, Garrett-Engele P, MacIsaac KD, Stevens RC, Sriram S, Chen R, et al. A quantitative atlas of polyadenylation in five mammals. Genome Research. 2012;**22**(6):1173-1183

[18] Lin Y, Li Z, Ozsolak F, Kim SW, Arango-Argoty G, Liu TT, et al. An

in-depth map of polyadenylation sites in cancer. Nucleic Acids Research. 2012;**40**(17):8460-8471

[19] Tian B, Manley JL. Alternative polyadenylation of mRNA precursors. Nature Reviews Molecular Cell Biology. 2017;**18**(1):18

[20] Danckwardt S, Hentze MW, Kulozik AE. 3′ end mRNA processing: Molecular mechanisms and implications for health and disease. The EMBO Journal. 2008;**27**(3):482-498

[21] Wickramasinghe VO, Laskey RA. Control of mammalian gene expression by selective mRNA export. Nature Reviews Molecular Cell Biology. 2015; **16**(7):431-442

[22] Lei EP, Krebber H, Silver PA. Messenger RNAs are recruited for nuclear export during transcription. Genes & Development. 2001;**15**(14):1771-1782

[23] Maniatis T, Reed R. An extensive network of coupling among gene expression machines. Nature. 2002; **416**(6880):499

[24] Neugebauer KM. On the importance of being co-transcriptional. Journal of Cell Science. 2002;**115**(20):3865-3871

[25] Hilleren P, McCarthy T, Rosbash M, Parker R, Jensen TH. Quality control of mRNA 3′-end processing is linked to the nuclear exosome. Nature. 2001;**413**(6855):538

[26] Hammell CM, Gross S, Zenklusen D, Heath CV, Stutz F, Moore C, et al. Coupling of termination, 3′ processing, and mRNA export. Molecular and Cellular Biology. 2002;**22**(18):6441-6457

[27] Kessler MM, Henry MF, Shen E, Zhao J, Gross S, Silver PA, et al. Hrp1, a sequence specific RNA-binding protein that shuttles between the nucleus and the cytoplasm, is required for mRNA 3′-end formation in yeast. Genes & Development. 1997;**11**(19):2545-2556

[28] Hector RE, Nykamp KR, Dheur S, Anderson JT, Non PJ, Urbinati CR, et al. Dual requirement for yeast hnRNP Nab2p in mRNA poly (A) tail length control and nuclear export. The EMBO Journal. 2002;**21**(7):1800-1810

[29] Hollams EM, Giles KM, Thomson AM, Leedman PJ. MRNA stability and the control of gene expression: Implications for human disease. Neurochemical Research. 2002;**27**(10):957-980

[30] Balagopal V, Fluch L, Nissan T. Ways and means of eukaryotic mRNA decay. Biochimica et Biophysica Acta (BBA)-Gene Regulatory Mechanisms. 2012;**1819**(6):593-603

[31] Adjibade P, Mazroui R. Control of mRNA turnover: Implication of cytoplasmic RNA granules. Seminars in Cell & Developmental Biology. 2014;**34**:15-23

[32] Yang E, van Nimwegen E, Zavolan M, Rajewsky N, Schroeder M, Magnasco M, et al. Decay rates of human mRNAs: Correlation with functional characteristics and sequence attributes. Genome Research. 2003;**13**(8):1863-1872

[33] Hollis GF, Gazdar AF, Bertness V, Kirsch IR. Complex translocation disrupts c-myc regulation in a human plasma cell myeloma. Molecular and Cellular Biology. 1988;**8**(1):124-129

[34] Aghib DF, Bishop JM, Ottolenghi S, Guerrasio A, Serra A, Saglio G. A 3'truncation of MYC caused by chromosomal translocation in a human T-cell leukemia increases mRNA stability. Oncogene. 1990;**5**(5):707-711

[35] Schuman EM, Dynes JL, Steward O. Synaptic regulation of translation of dendritic mRNAs. Journal of Neuroscience. 2006;**26**(27):7143-7146

[36] Thompson B, Wickens M, Kimble J. 19 translational control in development. Cold Spring Harbor Monograph Archive. 2007;**48**:507-544

[37] Morley SJ, Coldwell MJ. 16 matters of life and death: Translation initiation during apoptosis. Cold Spring Harbor Monograph Archive. 2007;**48**:433-458

[38] Holcik M, Sonenberg N. Translational control in stress and apoptosis. Nature Reviews Molecular Cell Biology. 2005; **6**(4):318

[39] Schneider RJ, Sonenberg N. 15 translational control in cancer development and progression. Cold Spring Harbor Monograph Archive. 2007;**48**:401-431

[40] Gebauer F, Hentze MW. Molecular mechanisms of translational control. Nature Reviews Molecular Cell Biology. 2004;**5**(10):827

[41] Mathews MB, Sonenberg N, Hershey JW. 1 origins and principles of translational control. Cold Spring Harbor Monograph Archive. 2007;**48**:1-40

[42] Preiss T, W. Hentze M. Starting the protein synthesis machine: Eukaryotic translation initiation. BioEssays. 2003;**25**(12):1201-1211

[43] Bashkirov VI, Scherthan H, Solinger JA, Buerstedde JM, Heyer WD. A mouse cytoplasmic exoribonuclease (mXRN1p) with preference for G4 tetraplex substrates. The Journal of Cell Biology. 1997;**136**(4):761-773

[44] Cougot N, Babajko S, Seraphin B. Cytoplasmic foci are sites of mRNA decay in human cells. The Journal of Cell Biology. 2004;**165**:31-40

[45] Ingelfinger D, Arndt-Jovin DJ, Lührmann R, Achsel T. The human LSm1-7 proteins colocalize with the mRNA-degrading enzymes Dcp1/2 and Xrn1 in distinct cytoplasmic foci. RNA. 2002;**8**(12):1489-1501

[46] Lykke-Andersen J. Identification of a human decapping complex associated with hUpf proteins in nonsense-mediated decay. Molecular and Cellular Biology. 2002;**22**(23):8114-8121

[47] Sheth U, Parker R. Decapping and decay of messenger RNA occur in cytoplasmic processing bodies. Science. 2003;**300**(5620):805-808

[48] Van Dijk E, Cougot N, Meyer S, Babajko S, Wahle E, Séraphin B. Human Dcp2: A catalytically active mRNA decapping enzyme located in specific cytoplasmic structures. The EMBO Journal. 2002;**21**(24):6915-6924

[49] Coller J, Parker R. General translational repression by activators of mRNA decapping. Cell. 2005;**122**(6): 875-886

[50] Brengues M, Teixeira D, Parker R. Movement of eukaryotic mRNAs between polysomes and cytoplasmic processing bodies. Science. 2005;**310**(5747):486-489

[51] Bartel DP. MicroRNAs: Genomics, biogenesis, mechanism, and function. Cell. 2004;**116**(2):281-297

[52] Filipowicz W. RNAi: The nuts and bolts of the RISC machine. Cell. 2005;**122**(1):17-20

[53] Sonenberg N, Hinnebusch AG. Regulation of translation initiation in eukaryotes: Mechanisms and biological targets. Cell. 2009;**136**(4):731-745

[54] Orphanides G, Lagrange T, Reinberg D. The general transcription factors of RNA Pol II. Genes & Development. 1996;**10**(21):2657-2683

[55] Sikorski TW, Buratowski S. The basal initiation machinery: Beyond the general transcription factors. Current Opinion in Cell Biology. 2009;**21**(3):344-351

[56] van Bakel H. Interactions of transcription factors with chromatin. In: A Handbook of Transcription Factors. Dordrecht: Springer; 2011. pp. 223-259

[57] Kornberg RD. Mediator and the mechanism of transcriptional activation. Trends in Biochemical Sciences. 2005;**30**(5):235-239

[58] Malik S, Roeder RG. The metazoan mediator co-activator complex as an integrative hub for transcriptional regulation. Nature Reviews Genetics. 2010;**11**(11):761

[59] Cevher MA, Shi Y, Li D, Chait BT, Malik S, Roeder RG. Reconstitution of active human core mediator complex reveals a critical role of the MED14 subunit. Nature Structural & Molecular Biology. 2014;**21**(12):1028

[60] Plaschka C, Lariviere L, Wenzeck L, Seizl M, Hemann M, Tegunov D, et al. Architecture of the RNA Pol II—Mediator core initiation complex. Nature. 2015;**518**(7539):376

[61] Harper TM, Taatjes DJ. The complex structure and function of mediator. Journal of Biological Chemistry. 2018;**293**(36):13778-13785

[62] Soutourina J. Transcription regulation by the mediator complex. Nature Reviews Molecular Cell Biology. 2018;**19**(4):262

[63] Kim YJ, Björklund S, Li Y, Sayre MH, Kornberg RD. A multiprotein mediator of transcriptional activation and its interaction with the C-terminal repeat domain of RNA Pol II. Cell. 1994;**77**(4):599-608

[64] Gaillard H, Tous C, Botet J, González-Aguilera C, Quintero MJ, Viladevall L, et al. Genome-wide analysis of factors affecting transcription elongation and DNA repair: A new role for PAF and Ccr4-not in transcription-coupled repair. PLoS Genetics. 2009;**5**(2):e1000364

[65] Kremer SB, Kim S, Jeon JO, Moustafa YW, Chen A, Zhao J, et al. Role of mediator in regulating Pol II elongation and nucleosome displacement in Saccharomyces cerevisiae. Genetics. 2012;**191**(1):95-106

[66] Mukundan B, Ansari A. Novel role for mediator complex subunit Srb5/Med18 in termination of transcription. Journal of Biological Chemistry. 2011;**286**(43):37053-37057

[67] Schneider M, Hellerschmied D, Schubert T, Amlacher S, Vinayachandran V, Reja R, et al. The nuclear pore-associated TREX-2 complex employs mediator to regulate gene expression. Cell. 2015;**162**(5):1016-1028

[68] Eyboulet F, Cibot C, Eychenne T, Neil H, Alibert O, Werner M, et al. Mediator links transcription and DNA repair by facilitating Rad2/XPG recruitment. Genes & Development. 2013;**27**(23):2549-2562

[69] Mehta S, Miklos I, Sipiczki M, Sengupta S, Sharma N. The Med8 mediator subunit interacts with the Rpb4 subunit of RNA Pol II and Ace2 transcriptional activator in Schizosaccharomyces pombe. FEBS Letters. 2009;**583**(19):3115-3120

[70] Bushnell DA, Westover KD, Davis RE, Kornberg RD. Structural basis

of transcription: An RNA Pol II-TFIIB cocrystal at 4.5 angstroms. Science. 2004;**303**(5660):983-988

[71] Kwak H, Lis JT. Control of transcriptional elongation. Annual Review of Genetics. 2013;**47**:483-508

[72] Laybourn PJ, Dahmus ME. Transcription-dependent structural changes in the C-terminal domain of mammalian RNA polymerase subunit IIa/o. Journal of Biological Chemistry. 1989;**264**(12):6693-6698

[73] Lu H, Flores O, Weinmann R, Reinberg D. The nonphosphorylated form of RNA Pol II preferentially associates with the preinitiation complex. Proceedings of the National Academy of Sciences. 1991;**88**(22):10004-10008

[74] Christmann JL, Dahmus ME. Monoclonal antibody specific for calf thymus RNA polymerases IIO and IIA. Journal of Biological Chemistry. 1981;**256**(22):11798-11803

[75] Jones JC, Phatnani HP, Haystead TA, MacDonald JA, Alam SM, Greenleaf AL. C-terminal repeat domain kinase I phosphorylates Ser2 and Ser5 of RNA Pol II C-terminal domain repeats. Journal of Biological Chemistry. 2004;**279**(24):24957-24964

[76] Allen BL, Taatjes DJ. The mediator complex: A central integrator of transcription. Nature Reviews Molecular Cell Biology. 2015;**16**(3):155

[77] Marshall NF, Price DH. Purification of P-TEFb, a transcription factor required for the transition into productive elongation. Journal of Biological Chemistry. 1995;**270**(21):12335-12338

[78] Marshall NF, Peng J, Xie Z, Price DH. Control of RNA Pol II elongation potential by a novel carboxyl-terminal domain kinase. Journal of Biological Chemistry. 1996;**271**(43):27176-27183

[79] Zhou Q, Li T, Price DH. RNA Pol II elongation control. Annual Review of Biochemistry. 2012;**81**:119-143

[80] Mayer A, Lidschreiber M, Siebert M, Leike K, Söding J, Cramer P. Uniform transitions of the general RNA Pol II transcription complex. Nature Structural & Molecular Biology. 2010;**17**(10): 1272-1278

[81] Booth GT, Wang IX, Cheung VG, Lis JT. Divergence of a conserved elongation factor and transcription regulation in budding and fission yeast. Genome Research. 2016;**26**(6):799-811

[82] Conaway JW, Shilatifard A, Dvir A, Conaway RC. Control of elongation by RNA Pol II. Trends in Biochemical Sciences. 2000;**25**(8):375-380

[83] Ucker DS, Yamamoto KR. Early events in the stimulation of mammary tumor virus RNA synthesis by glucocorticoids. Novel assays of transcription rates. Journal of Biological Chemistry. 1984;**259**(12):7416-7420

[84] Thummel CS, Burtis KC, Hogness DS. Spatial and temporal patterns of E74 transcription during Drosophila development. Cell. 1990;**61**(1):101-111

[85] Tennyson CN, Klamut HJ, Worton RG. The human dystrophin gene requires 16 hours to be transcribed and is cotranscriptionally spliced. Nature Genetics. 1995;**9**(2):184

[86] Reines D, Conaway JW, Conaway RC. The RNA Pol II general elongation factors. Trends in Biochemical Sciences. 1996;**21**(9):351-355

[87] Uptain SM, Kane CM, Chamberlin MJ. Basic mechanisms of transcript elongation

and its regulation. Annual Review of Biochemistry. 1997;**66**(1):117-172

[88] Wada T, Takagi T, Yamaguchi Y, Ferdous A, Imai T, Hirose S, et al. DSIF, a novel transcription elongation factor that regulates RNA Pol II processivity, is composed of human Spt4 and Spt5 homologs. Genes & Development. 1998;**12**(3):343-356

[89] Hartzog GA, Wada T, Handa H, Winston F. Evidence that Spt4, Spt5, and Spt6 control transcription elongation by RNA Pol II inSaccharomyces cerevisiae. Genes & Development. 1998;**12**(3): 357-369

[90] Orphanides G, Reinberg D. RNA Pol II elongation through chromatin. Nature. 2000;**407**(6803):471

[91] Arndt KM, Kane CM. Running with RNA polymerase: Eukaryotic transcript elongation. Trends in Genetics. 2003;**19**(10):543-550

[92] Bradsher JN, Jackson KW, Conaway RC, Conaway JW. RNA Pol II transcription factor SIII. I. Identification, purification, and properties. Journal of Biological Chemistry. 1993;**268**(34): 25587-25593

[93] Garrett KP, Aso T, Bradsher JN, Foundling SI, Lane WS, Conaway RC, et al. Positive regulation of general transcription factor SIII by a tailed ubiquitin homolog. Proceedings of the National Academy of Sciences. 1995;**92**(16):7172-7176

[94] Shilatifard A, Lane WS, Jackson KW, Conaway RC, Conaway JW. An RNA Pol II elongation factor encoded by the human ELL gene. Science. 1996;**271**(5257):1873-1876

[95] Shilatifard A, Haque D, Conaway RC, Conaway JW. Structure and function of RNA Pol II elongation factor ELL identification of two overlapping ell functional domains that govern its interaction with polymerase and the ternary elongation complex. Journal of Biological Chemistry. 1997;**272**(35):22355-22363

[96] Shilatifard A, Duan DR, Haque D, Florence C, Schubach WH, Conaway JW, et al. ELL2, a new member of an ELL family of RNA Pol II elongation factors. Proceedings of the National Academy of Sciences. 1997;**94**(8):3639-3643

[97] Yamaguchi Y, Wada T, Watanabe D, Takagi T, Hasegawa J, Handa H. Structure and function of the human transcription elongation factor DSIF. Journal of Biological Chemistry. 1999;**274**(12):8085-8092

[98] Yan Q, Moreland RJ, Conaway JW, Conaway RC. Dual roles for transcription factor IIF in promoter escape by RNA Pol II. Journal of Biological Chemistry. 1999;**274**(50):35668-35675

[99] Conaway RC, Conaway JW. General initiation factors for RNA Pol II. Annual Review of Biochemistry. 1993;**62**(1):161-190

[100] Tan S, Aso T, Conaway RC, Conaway JW. Roles for both the RAP30 and RAP74 subunits of transcription factor IIF in transcription initiation and elongation by RNA Pol II. Journal of Biological Chemistry. 1994;**269**(41):25684-25691

[101] Kephart DD, Wang BQ, Burton ZF, Price DH. Functional analysis of Drosophila factor 5 (TFIIF), a general transcription factor. Journal of Biological Chemistry. 1994;**269**(18):13536-13543

[102] Wang BQ, Burton ZF. Functional domains of human RAP74 including a masked polymerase binding domain.

Journal of Biological Chemistry. 1995;**270**(45):27035-27044

[103] Lee TI, Young RA. Transcription of eukaryotic protein-coding genes. Annual Review of Genetics. 2000;**34**(1):77-137

[104] Goodrich JA, Tjian R. Transcription factors IIE and IIH and ATP hydrolysis direct promoter clearance by RNA Pol II. Cell. 1994;**77**(1):145-156

[105] Ohkuma Y, Roeder RG. Regulation of TFIIH ATPase and kinase activities by TFIIE during active initiation complex formation. Nature. 1994;**368**(6467):160

[106] Holstege FC, Van der Vliet PC, Timmers HT. Opening of an RNA Pol II promoter occurs in two distinct steps and requires the basal transcription factors IIE and IIH. The EMBO Journal. 1996;**15**(7):1666-1677

[107] Schilbach S, Hantsche M, Tegunov D, Dienemann C, Wigge C, Urlaub H, et al. Structures of transcription pre-initiation complex with TFIIH and mediator. Nature. 2017;**551**(7679):204

[108] Glover-Cutter K, Larochelle S, Erickson B, Zhang C, Shokat K, Fisher RP, et al. TFIIHassociated Cdk7 kinase functions in phosphorylation of C-terminal domain Ser7 residues, promoter-proximal pausing, and termination by RNA Pol II. Molecular and Cellular Biology. 2009;**29**(20):5455-5464

[109] Mourgues S, Gautier V, Lagarou A, Bordier C, Mourcet A, Slingerland J, et al. ELL, a novel TFIIH partner, is involved in transcription restart after DNA repair. Proceedings of the National Academy of Sciences. 2013;**110**(44):17927-17932

[110] Dubois MF, Nguyen VT, Bellier S, Bensaude O. Inhibitors of transcription such as 5, 6-dichloro- 1-beta-D-ribofuranosylbenzimidazole and isoquinoline sulfonamide derivatives (H-8 and H-7) promote dephosphorylation of the carboxyl-terminal domain of RNA Pol II largest subunit. Journal of Biological Chemistry. 1994;**269**(18):13331-13336

[111] Mancebo HS, Lee G, Flygare J, Tomassini J, Luu P, Zhu Y, et al. P-TEFb kinase is required for HIV tat transcriptional activation in vivo and in vitro. Genes & Development. 1997;**11**(20):2633-2644

[112] Zhu Y, Pe'ery T, Peng J, Ramanathan Y, Marshall N, Marshall T, et al. Transcription elongation factor P-TEFb is required for HIV-1 tat transactivation in vitro. Genes & Ddevelopment. 1997;**11**(20):2622-2632

[113] Swanson MS, Malone EA, Winston FR. SPT5, an essential gene important for normal transcription in Saccharomyces cerevisiae, encodes an acidic nuclear protein with a carboxy-terminal repeat. Molecular and Cellular Biology. 1991;**11**(6):3009-3019

[114] Hartzog GA, Basrai MA, Ricupero-Hovasse SL, Hieter P, Winston F. Identification and analysis of a functional human homolog of the SPT4 gene of Saccharomyces cerevisiae. Molecular and Cellular Biology. 1996;**16**(6):2848-2856

[115] Larochelle S, Batliner J, Gamble MJ, Barboza NM, Kraybill BC, Blethrow JD, et al. Dichotomous but stringent substrate selection by the dual-function Cdk7 complex revealed by chemical genetics. Nature Structural & Molecular Biology. 2006;**13**(1):55

[116] Patel SA, Simon MC. Functional analysis of the CDK7. Cyclin H. Mat1 complex in mouse embryonic stem cells and embryos. Journal of Biological Chemistry. 2010;**285**(20):15587-15598

[117] Yamamoto J, Hagiwara Y, Chiba K, Isobe T, Narita T, Handa H, et al. DSIF

and NELF interact with integrator to specify the correct post-transcriptional fate of snRNA genes. Nature Communications. 2014;**5**:4263

[118] Yamaguchi Y, Filipovska J, Yano K, Furuya A, Inukai N, Narita T, et al. Stimulation of RNA Pol II elongation by hepatitis delta antigen. Science. 2001;**293**(5527):124-127

[119] Sims RJ, Belotserkovskaya R, Reinberg D. Elongation by RNA Pol II: The short and long of it. Genes & Development. 2004;**18**(20):2437-2468

[120] Yamaguchi Y, Shibata H, Handa H. Transcription elongation factors DSIF and NELF: Promoter proximal pausing and beyond. Biochimica Et Biophysica Acta (BBA)-Gene Regulatory Mechanisms. 2013;**1829**(1):98-104

[121] Pei Y, Shuman S. Interactions between fission yeast mRNA capping enzymes and elongation factor Spt5. Journal of Biological Chemistry. 2002;**277**(22):19639-19648

[122] Ivanov D, Kwak YT, Guo J, Gaynor RB. Domains in the SPT5 protein that modulate its transcriptional regulatory properties. Molecular and Cellular Biology. 2000;**20**(9):2970-2983

[123] Kim JB, Sharp PA. Positive transcription elongation factor B phosphorylates hSPT5 and RNA Pol II carboxyl-terminal domain independently of cyclin-dependent kinaseactivating kinase. Journal of Biological Chemistry. 2001;**276**(15):12317-12323

[124] Shilatifard A, Conaway RC, Conaway JW. The RNA Pol II elongation complex. Annual Review of Biochemistry. 2003;**72**(1):693-715

[125] Wada T, Orphanides G, Hasegawa J, Kim DK, Shima D, Yamaguchi Y, et al. FACT relieves DSIF/NELF-mediated inhibition of transcriptional elongation and reveals functional differences between P-TEFb and TFIIH. Molecular Cell. 2000;**5**(6):1067-1072

[126] Rice AP. Dysregulation of positive transcription elongation factor B and myocardial hypertrophy. Circulation Research. 2009;**104**(12):1327-1329

[127] Peterlin BM, Price DH. Controlling the elongation phase of transcription with P-TEFb. Molecular Cell. 2006;**23**(3): 297-305

[128] Jang MK, Mochizuki K, Zhou M, Jeong HS, Brady JN, Ozato K. The bromodomain protein Brd4 is a positive regulatory component of P-TEFb and stimulates RNA Pol II dependent transcription. Molecular Cell. 2005;**19**(4):523-534

[129] Yang Z, Yik JH, Chen R, He N, Jang MK, Ozato K, et al. Recruitment of P-TEFb for stimulation of transcriptional elongation by the bromodomain protein Brd4. Molecular Cell. 2005;**19**(4):535-545

[130] Smith E, Lin C, Shilatifard A. The super elongation complex (SEC) and MLL in development and disease. Genes & Development. 2011;**25**(7):661-672

[131] Yamamoto M, Onogi H, Kii I, Yoshida S, Iida K, Sakai H, et al. CDK9 inhibitor FIT-039 prevents replication of multiple DNA viruses. The Journal of Clinical Investigation. 2014;**124**(8) :3479-3488

[132] Hu Z, Yik JH, Cissell DD, Michelier PV, Athanasiou KA, Haudenschild DR. Inhibition of CDK9 prevents mechanical injury-induced inflammation, apoptosis and matrix degradation in cartilage explants. European Cells & Materials. 2016;**30**:200

[133] Tanaka T, Okuyama-Dobashi K, Murakami S, Chen W, Okamoto T, Ueda K, et al. Inhibitory effect of CDK9 inhibitor FIT-039 on hepatitis B virus propagation. Antiviral Research. 2016;**133**:156-164

[134] Zaborowska J, Isa NF, Murphy S. P-TEFb goes viral. BioEssays. 2016;**38**: S75-S85

[135] Zhang J, Li G, Ye X. Cyclin T1/ CDK9 interacts with influenza A virus polymerase and facilitates its association with cellular RNA Pol II. Journal of Virology. 2010;**84**(24):12619-12627

[136] Schrecengost RS, Green CL, Zhuang Y, Keller SN, Smith RA, Maines LW, et al. In vitro and in vivo antitumor and anti-inflammatory capabilities of the novel GSK3 and CDK9 inhibitor ABC1183. Journal of Pharmacology and Experimental Therapeutics. 2018;**365**(1):107-116

[137] Qing Y, Wang X, Wang H, Hu P, Li H, Yu X, et al. Pharmacologic targeting of the P-TEFb complex as a therapeutic strategy for chronic myeloid leukemia. Cell Communication and Signaling. 2021;**19**(1):1-6

[138] Erwin GS, Grieshop MP, Ali A, Qi J, Lawlor M, Kumar D, et al. Synthetic transcription elongation factors license transcription across repressive chromatin. Science. 2017;**358**(6370):1617-1622

[139] Conaway JW, Conaway RC. Transcription elongation and human disease. Annual Review of Biochemistry. 1999;**68**(1):301-319

[140] Sharma N. Regulation of RNA Pol II-mediated transcriptional elongation: Implications in human disease. IUBMB Life. 2016;**68**(9):709-716

[141] Gerber M, Eissenberg JC, Kong S, Tenney K, Conaway JW, Conaway RC, et al. In vivo requirement of the RNA Pol II elongation factor elongin A for proper gene expression and development. Molecular and Cellular Biology. 2004;**24**(22):9911-9919

[142] Schoenfeld AR, Davidowitz EJ, Burk RD. Elongin BC complex prevents degradation of von Hippel-Lindau tumor suppressor gene products. Proceedings of the National Academy of Sciences. 2000;**97**(15):8507-8512

[143] Yasukawa T, Kamura T, Kitajima S, Conaway RC, Conaway JW, Aso T. Mammalian Elongin A complex mediates DNA-damage-induced ubiquitylation and degradation of Rpb1. The EMBO Journal. 2008;**27**(24): 3256-3266

[144] Harreman M, Taschner M, Sigurdsson S, Anindya R, Reid J, Somesh B, et al. Distinct ubiquitin ligases act sequentially for RNA Pol II polyubiquitylation. Proceedings of the National Academy of Sciences. 2009;**106**(49):20705-20710

[145] Nance MA, Berry SA. Cockayne syndrome: Review of 140 cases. American Journal of Medical Genetics. 1992;**42**(1):68-84

[146] Wilson BT, Stark Z, Sutton RE, Danda S, Ekbote AV, Elsayed SM, et al. The Cockayne syndrome natural history (CoSyNH) study: Clinical findings in 102 individuals and recommendations for care. Genetics in Medicine. 2016;**18**(5):483

[147] Venema JL, Mullenders LH, Natarajan AT, Van Zeeland AV, Mayne LV. The genetic defect in Cockayne syndrome is associated with a defect in repair of UV-induced DNA damage in transcriptionally active DNA. Proceedings of the National Academy of

Sciences of the United States of America. 1990;**87**(12):4707

[148] van Hoffen A, Natarajan AT, Mayne LV, Zeeland AA, Mullenders LH, Venema J. Deficient repair of the transcribed strand of active genes in Cockayne's syndrome cells. Nucleic Acids Research. 1993;**21**(25):5890-5895

[149] Vélez-Cruz R, Egly JM. Cockayne syndrome group B (CSB) protein: At the crossroads of transcriptional networks. Mechanisms of Ageing and Development. 2013;**134**(5-6):234-242

[150] Mellon I, Spivak G, Hanawalt PC. Selective removal of transcription-blocking DNA damage from the transcribed strand of the mammalian DHFR gene. Cell. 1987;**51**(2):241-249

[151] Troelstra C, van Gool A, de Wit J, Vermeulen W, Bootsma D, Hoeijmakers JH. ERCC6, a member of a subfamily of putative helicases, is involved in Cockayne's syndrome and preferential repair of active genes. Cell. 1992;**71**(6): 939-953

[152] Hanawalt PC. Transcription-coupled repair and human disease. Science. 1994;**266**(5193):1957

[153] Hanawalt PC, Spivak G. Transcription-coupled DNA repair: Two decades of progress and surprises. Nature Reviews Molecular Cell Biology. 2008;**9**(12):958-970

[154] Mason PB, Struhl K. Distinction and relationship between elongation rate and processivity of RNA Pol II in vivo. Molecular Cell. 2005;**17**(6):831-840

[155] Ishibashi T, Dangkulwanich M, Coello Y, Lionberger TA, Lubkowska L, Ponticelli AS, et al. Transcription factors IIS and IIF enhance transcription efficiency by differentially modifying RNA polymerase pausing dynamics. Proceedings of the National Academy of Sciences. 2014;**111**(9):3419-3424

[156] Ito T, Saso K, Arimitsu N, Sekimizu K. Defective FESTA/EAF2-mediated transcriptional activation in S-II-deficient embryonic stem cells. Biochemical and Biophysical Research Communications. 2007;**363**(3):603-609

[157] Thirman MJ, Levitan DA, Kobayashi H, Simon MC, Rowley JD. Cloning of ELL, a gene that fuses to MLL in at (11; 19)(q23; p13. 1) in acute myeloid leukemia. Proceedings of the National Academy of Sciences. 1994;**91**(25):12110-12114

[158] Byun JS, Fufa TD, Wakano C, Fernandez A, Haggerty CM, Sung MH, et al. ELL facilitates RNA Pol II pause site entry and release. Nature Communications. 2012;**3**:633

[159] Liu L, Ai J, Xiao W, Liu J, Wang Y, Xin D, et al. ELL is an HIF-1 partner that regulates and responds to hypoxia response in PC3 cells. The Prostate. 2010;**70**(7):797-805

[160] Fufa TD, Byun JS, Wakano C, Fernandez AG, Pise-Masison CA, Gardner K. The tax oncogene enhances ELL incorporation into p300 and P-TEFb containing protein complexes to activate transcription. Biochemical and Biophysical Research Communications. 2015;**465**(1):5-11

[161] Martincic K, Alkan SA, Cheatle A, Borghesi L, Milcarek C. Transcription elongation factor ELL2 directs immunoglobulin secretion in plasma cells by stimulating altered RNA processing. Nature Immunology. 2009;**10**(10):1102

[162] Swaminathan B, Thorleifsson G, Jöud M, Ali M, Johnsson E, Ajore R,

et al. Variants in ELL2 influencing immunoglobulin levels associate with multiple myeloma. Nature Communications. 2015;**6**:7213

[163] Shell SA, Martincic K, Tran J, Milcarek C. Increased phosphorylation of the carboxyl-terminal domain of RNA Pol II and loading of polyadenylation and cotranscriptional factors contribute to regulation of the ig heavy chain mRNA in plasma cells. The Journal of Immunology. 2007;**179**(11):7663-7673

[164] Pascal LE, Masoodi KZ, Liu J, Qiu X, Song Q, Wang Y, et al. Conditional deletion of ELL2 induces murine prostate intraepithelial neoplasia. Journal of Endocrinology. 2017;**235**(2):123-136

[165] Ali M, Ajore R, Wihlborg AK, Niroula A, Swaminathan B, Johnsson E, et al. The multiple myeloma risk allele at 5q15 lowers ELL2 expression and increases ribosomal gene expression. Nature Communications. 2018;**9**(1):1649

[166] Ahn HJ, Kim G, Park KS. Ell3 stimulates proliferation, drug resistance, and cancer stem cell properties of breast cancer cells via a MEK/ERK-dependent signaling pathway. Biochemical and Biophysical Research Communications. 2013;**437**(4):557-564

[167] Alexander LE, Watters J, Reusch JA, Maurin M, Nepon-Sixt BS, Vrzalikova K, et al. Selective expression of the transcription elongation factor ELL3 in B cells prior to ELL2 drives proliferation and survival. Molecular Immunology. 2017;**91**:8-16

[168] Simone F, Polak PE, Kaberlein JJ, Luo RT, Levitan DA, Thirman MJ. EAF1, a novel ELLassociated factor that is delocalized by expression of the MLL-ELL fusion protein. Blood. 2001;**98**(1):201-209

[169] Simone F, Luo RT, Polak PE, Kaberlein JJ, Thirman MJ. ELL-associated factor 2 (EAF2), a functional homolog of EAF1 with alternative ELL binding properties. Blood. 2003;**101**(6):2355-2362

[170] Hahn J, Xiao W, Jiang F, Simone F, Thirman MJ, Wang Z. Apoptosis induction and growth suppression by U19/Eaf2 is mediated through its ELL-binding domain. The Prostate. 2007;**67**(2):146-153

[171] Kong SE, Banks CA, Shilatifard A, Conaway JW, Conaway RC. ELL-associated factors 1 and 2 are positive regulators of RNA Pol II elongation factor ELL. Proceedings of the National Academy of Sciences. 2005;**102**(29):10094-10098

[172] Luo RT, Lavau C, Du C, Simone F, Polak PE, Kawamata S, et al. The elongation domain of ELL is dispensable but its ELL-associated factor 1 interaction domain is essential for MLL-ELL-induced leukemogenesis. Molecular and Cellular Biology. 2001;**21**(16):5678-5687

[173] Qiao Z, Wang D, Hahn J, Ai J, Wang Z. Pirin down-regulates the EAF 2/U 19 protein and alleviates its growth inhibition in prostate cancer cells. The Prostate. 2014;**74**(2):113-120

[174] Zhang Q, Zeng L, Zhao C, Ju Y, Konuma T, Zhou MM. Structural insights into histone crotonyllysine recognition by the AF9 YEATS domain. Structure. 2016;**24**(9):1606-1612

[175] Wang Y, Pascal LE, Zhong M, Ai J, Wang D, Jing Y, et al. Combined loss of EAF2 and p53 induces prostate carcinogenesis in male mice. Endocrinology. 2017;**158**(12):4189-4205

[176] Yang T, Jing Y, Dong J, Yu X, Zhong M, Pascal LE, et al. Regulation of ELL2 stability and polyubiquitination by EAF2 in prostate cancer cells. The Prostate. 2018;**78**(15):1201-1212

[177] Zhong M, Pascal LE, Cheng E, Masoodi KZ, Chen W, Green A, et al.

Concurrent EAF2 and ELL2 loss phenocopies individual EAF2 or ELL2 loss in prostate cancer cells and murine prostate. American Journal of Clinical and Experimental Urology. 2018;**6**(6):234

[178] Xiao W, Zhang Q, Habermacher G, Yang X, Zhang AY, Cai X, et al. U19/Eaf2 knockout causes lung adenocarcinoma, B-cell lymphoma, hepatocellular carcinoma and prostatic intraepithelial neoplasia. Oncogene. 2008;**27**(11):1536

[179] Xiao W, Ai J, Habermacher G, Volpert O, Yang X, Zhang AY, et al. U19/Eaf2 binds to and stabilizes von hippel-Lindau protein. Cancer Research. 2009;**69**(6):2599-2606

[180] Li Y, Sabari BR, Panchenko T, Wen H, Zhao D, Guan H, et al. Molecular coupling of histone crotonylation and active transcription by AF9 YEATS domain. Molecular Cell. 2016;**62**(2):181-193

[181] Jo YS, Kim SS, Kim MS, Yoo NJ, Lee SH. Candidate tumor suppressor gene EAF2 is mutated in colorectal and gastric cancers. Pathology Oncology Research. 2019;**25**(2):823-824

[182] Pascal LE, Su F, Wang D, Ai J, Song Q, Wang Y, et al. Conditional deletion of eaf1 induces murine prostatic intraepithelial neoplasia in mice. Neoplasia. 2019;**21**(8):752-764

[183] Lin C, Smith ER, Takahashi H, Lai KC, Martin-Brown S, Florens L, et al. AFF4, a component of the ELL/P-TEFb elongation complex and a shared subunit of MLL chimeras, can link transcription elongation to leukemia. Molecular Cell. 2010;**37**(3):429-437

[184] Hu D, Smith ER, Garruss AS, Mohaghegh N, Varberg JM, Lin C, et al. The little elongation complex functions at initiation and elongation phases of snRNA gene transcription. Molecular Cell. 2013;**51**(4):493-505

[185] Lu H, Li Z, Xue Y, Schulze-Gahmen U, Johnson JR, Krogan NJ, et al. AFF1 is a ubiquitous P-TEFb partner to enable tat extraction of P-TEFb from 7SK snRNP and formation of SECs for HIV transactivation. Proceedings of the National Academy of Sciences. 2014;**111**(1):E15-E24

[186] Dahlberg O, Shilkova O, Tang M, Holmqvist PH, Mannervik M. P-TEFb, the super elongation complex and mediator regulate a subset of non-paused genes during early Drosophila embryo development. PLoS Genetics. 2015;**11**(2):e1004971

[187] Lin C, Garrett AS, De Kumar B, Smith ER, Gogol M, Seidel C, et al. Dynamic transcriptional events in embryonic stem cells mediated by the super elongation complex (SEC). Genes & Development. 2011;**25**(14):1486-1498

[188] Gopalan S, Gibbon DM, Banks CA, Zhang Y, Florens LA, Washburn MP, et al. Schizosaccharomyces pombe Pol II transcription elongation factor ELL functions as part of a rudimentary super elongation complex. Nucleic Acids Research. 2018;**46**(19):10095-10105

[189] He N, Liu M, Hsu J, Xue Y, Chou S, Burlingame A, et al. HIV-1 tat and host AFF4 recruit two transcription elongation factors into a bifunctional complex for coordinated activation of HIV-1 transcription. Molecular Cell. 2010;**38**(3):428-438

[190] Sobhian B, Laguette N, Yatim A, Nakamura M, Levy Y, Kiernan R, et al. HIV-1 tat assembles a multifunctional transcription elongation complex and stably associates with the 7SK snRNP. Molecular Cell. 2010;**38**(3):439-451

[191] Yokoyama A, Lin M, Naresh A, Kitabayashi I, Cleary ML. A higher-order complex containing AF4 and ENL family proteins with P-TEFb facilitates oncogenic and physiologic MLL-dependent transcription. Cancer Cell. 2010;**17**(2):198-212

[192] Luo Z, Lin C, Guest E, Garrett AS, Mohaghegh N, Swanson S, et al. The super elongation complex family of RNA pol II elongation factors: Gene target specificity and transcriptional output. Molecular and Cellular Biology. 2012;**32**(13):2608-2617

[193] Lu H, Li Z, Zhang W, Schulze-Gahmen U, Xue Y, Zhou Q. Gene target specificity of the super elongation complex (SEC) family: How HIV-1 tat employs selected SEC members to activate viral transcription. Nucleic Acids Research. 2015;**43**(12):5868-5879

[194] Izumi K, Nakato R, Zhang Z, Edmondson AC, Noon S, Dulik MC, et al. Germline gain-of-function mutations in AFF4 cause a developmental syndrome functionally linking the super elongation complex and cohesin. Nature Genetics. 2015;**47**(4):338-344

[195] Liang K, Smith ER, Aoi Y, Stoltz KL, Katagi H, Woodfin AR, et al. Targeting processive transcription elongation via SEC disruption for MYC-induced cancer therapy. Cell. 2018;**175**(3):766-779

[196] Kuehner JN, Pearson EL, Moore C. Unravelling the means to an end: RNA Pol II transcription termination. Nature Reviews Molecular Cell Biology. 2011;**12**(5):283-294

[197] Jensen TH, Jacquier A, Libri D. Dealing with pervasive transcription. Molecular Cell. 2013;**52**(4):473-484

[198] Porrua O, Libri D. Transcription termination and the control of the transcriptome: Why, where and how to stop. Nature Reviews Molecular Cell Biology. 2015;**16**(3):190-202

[199] Rosonina E, Kaneko S, Manley JL. Terminating the transcript: Breaking up is hard to do. Genes & Development. 2006;**20**(9):1050-1056

[200] Mischo HE, Gómez-González B, Grzechnik P, Rondón AG, Wei W, Steinmetz L, et al. Yeast Sen1 helicase protects the genome from transcription-associated instability. Molecular Cell. 2011;**41**(1):21-32

[201] Richard P, Manley JL. Transcription termination by nuclear RNA polymerases. Genes & Development. 2009;**23**(11):1247-1269

[202] Shin Y, Brangwynne CP. Liquid phase condensation in cell physiology and disease. Science. 2017;**357**(6357)

[203] Kwon I, Kato M, Xiang S, Wu L, Theodoropoulos P, Mirzaei H, et al. Phosphorylation-regulated binding of RNA Pol II to fibrous polymers of low-complexity domains. Cell. 2013;**155**(5):1049-1060

[204] Boehning M, Dugast-Darzacq C, Rankovic M, Hansen AS, Yu T, Marie-Nelly H, et al. RNA Pol II clustering through carboxy-terminal domain phase separation. Nature Structural & Molecular Biology. 2018;**25**(9):833-840

[205] Lu H, Yu D, Hansen AS, Ganguly S, Liu R, Heckert A, et al. Phase-separation mechanism for C-terminal hyper-phosphorylation of RNA Pol II. Nature. 2018;**558**(7709):318-323

[206] Guo C, Che Z, Yue J, Xie P, Hao S, Xie W, et al. ENL initiates multivalent phase separation of the super elongation complex (SEC) in controlling rapid transcriptional activation. Science Advances. 2020;**6**(14):eaay4858

[207] Wan L, Chong S, Xuan F, Liang A, Cui X, Gates L, et al. Impaired cell fate through gain-of-function mutations in a chromatin reader. Nature. 2020;**577**(7788):121-126

[208] Isoda T, Moore AJ, He Z, Chandra V, Aida M, Denholtz M, et al. Non-coding transcription instructs chromatin folding and compartmentalization to dictate enhancer-promoter communication and T cell fate. Cell. 2017;**171**(1):103-119

[209] Mumbach MR, Granja JM, Flynn RA, Roake CM, Satpathy AT, Rubin AJ, et al. HiChIRP reveals RNA-associated chromosome conformation. Nature Methods. 2019;**16**(6):489-492

[210] Luo S, Lu JY, Liu L, Yin Y, Chen C, Han X, et al. Divergent lncRNAs regulate gene expression and lineage differentiation in pluripotent cells. Cell Stem Cell. 2016;**18**(5):637-652

[211] Kim YJ, Xie P, Cao L, Zhang MQ, Kim TH. Global transcriptional activity dynamics reveal functional enhancer RNAs. Genome Research. 2018;**28**(12): 1799-1811

[212] Lee S, Kopp F, Chang TC, Sataluri A, Chen B, Sivakumar S, et al. Noncoding RNA NORAD regulates genomic stability by sequestering PUMILIO proteins. Cell. 2016;**164**(1-2):69-80

[213] Tichon A, Perry RB, Stojic L, Ulitsky I. SAM68 is required for regulation of Pumilio by the NORAD long noncoding RNA. Genes & Development. 2018;**32**(1):70-78

[214] Carrieri C, Cimatti L, Biagioli M, Beugnet A, Zucchelli S, Fedele S, et al. Long non-coding antisense RNA controls Uchl1 translation through an embedded SINEB2 repeat. Nature. 2012;**491**(7424):454-457

[215] Aguilo F, Zhou MM, Walsh MJ. Long noncoding RNA, polycomb, and the ghosts haunting INK4b-ARF-INK4a expression. Cancer Research. 2011;**71**(16):5365-5369

[216] Popov N, Gil J. Epigenetic regulation of the INK4b-ARF-INK4a locus: In sickness and in health. Epigenetics. 2010;**5**(8):685-690

[217] Kotake Y, Nakagawa T, Kitagawa K, Suzuki S, Liu N, Kitagawa M, et al. Long non-coding RNA ANRIL is required for the PRC2 recruitment to and silencing of p15 INK4B tumor suppressor gene. Oncogene. 2011;**30**(16):1956-1962

[218] Yap KL, Li S, Muñoz-Cabello AM, Raguz S, Zeng L, Mujtaba S, et al. Molecular interplay of the noncoding RNA ANRIL and methylated histone H3 lysine 27 by polycomb CBX7 in transcriptional silencing of INK4a. Molecular Cell. 2010;**38**(5):662-674

[219] Faghihi MA, Zhang M, Huang J, Modarresi F, Van der Brug MP, Nalls MA, et al. Evidence for natural antisense transcript-mediated inhibition of microRNA function. Genome Biology. 2010;**11**(5):1-3

[220] Feng L, Liao YT, He JC, Xie CL, Chen SY, Fan HH, et al. Plasma long non-coding RNA BACE1 as a novel biomarker for diagnosis of Alzheimer disease. BMC Neurology. 2018;**18**(1):1-8

[221] Perry RB, Hezroni H, Goldrich MJ, Ulitsky I. Regulation of neuroregeneration by long noncoding RNAs. Molecular Cell. 2018;**72**(3):553-567

Chapter 3

# Biophysical and Biochemical Approaches for R-Loop Sensing Mechanism

*Na Young Cheon, Subin Kim and Ja Yil Lee*

**Abstract**

An R-loop is a triple-stranded nucleic acid structure consisting of a DNA–RNA hybrid and a displaced single-stranded DNA. R-loops are associated with diverse biological reactions, such as immune responses and gene regulation, and dysregulated R-loops can cause genomic instability and replication stress. Therefore, investigating the formation, regulation, and elimination of R-loops is important for understanding the molecular mechanisms underlying biological processes and diseases related to R-loops. Existing research has primarily focused on R-loop detection. In this chapter, we introduce a variety of biochemical and biophysical techniques for R-loop sensing and visualization both *in vivo* and *in vitro*, including single-molecule imaging. These methods can be used to investigate molecular mechanisms underlying R-loop search and identification.

**Keywords:** R-loop, genetic instability, R-loop sensing, and single-molecule imaging

## 1. Introduction

### 1.1 History of R-loops

R-loops are three-stranded nucleic acid structures consisting of a DNA–RNA hybrid and a displaced single-stranded (ss) DNA. They were first described in 1976 by Thomas *et al.*, who observed a hybridized form of ribosomal RNA and ribosomal DNA of *Saccharomyces cerevisiae* 26S [1]. Structurally, RNA–DNA hybrids adopt an A-form structure [2, 3]. The structure of an R-loop formed with RNA polymerase or CRISPR-Cas9 shows that R-loops have a helical RNA–DNA hybrid structure and a dissociated ssDNA [4–7].

### 1.2 Biological functions of R-loops

RNA–DNA hybrids can be formed from GC-rich clusters during transcription or primer synthesis of DNA replication [2, 8]. Because nucleic acid strands are stabilized when they form a double-stranded conformation, the nascent RNA is hybridized to the template DNA strand when the double-stranded (ds)DNA is denatured during replication or transcription [2]. Therefore, R-loops can form at any time when there



IntechOpen

is a chance that RNA can be annealed with its template DNA. The thread-back model proposes that R-loop formation stems from the annealing of RNA with DNA when the DNA left behind the transcriptome is negatively supercoiled and unwound [9]. Previous studies support this model [10–13]. Incomplete transcription elongation and termination also induce RNA–DNA hybrid formation, and the denaturation of duplex DNA by negative supercoiling increases R-loop formation.

R-loops have multiple roles in diverse biological reactions (**Figure 1**). First, R-loops induce genetic rearrangements in B-cells during immunoglobulin class switching [14]. R-loop formation is promoted by transcription through switched immunoglobulin loci, and the R-loop provides a ssDNA substrate for activation-induced deaminase (AID), which converts cytosine to uracil in both DNA strands. Uracil is subsequently removed by uracil glycosylase, and apurinic/apyrimidinic endonuclease makes nicks at the abasic sites and induces DNA double-strand breaks. During DNA double-strand break repair, the immunoglobulin locus is rearranged to change the level of antibodies generated. R-loops can also regulate both the activation and termination of transcription. Most human promoters are associated with CpG islands [15]. Ginno *et al*. demonstrated that R-loops are located at promoter sites that have CpG islands and proposed that R-loops protect the template DNA strand from gene-silencing methylation [16]. R-loop stabilization at the promoter region also regulates transcription. Flowering Locus C (FLC) is a transcription repressor of *Arabidopsis thaliana* that is regulated by COOLAIR through antisense transcription. Sun *et al*. found that the homeodomain protein AtNDX stabilizes R-loops by binding their displaced ssDNA at the *COOLAIR* promoter, thus inhibiting *COOLAIR* transcription and regulating *FLC* expression [17]. Antisense transcription-mediated R-loop formation at the *Vimentin* (*VIM*) promoter induces local chromatin decondensation and enhances gene expression [18]. R-loops also play a role in chromatin organization. For example, they are tightly linked to H3 S10 phosphorylation, which is a mark of chromatin condensation [19]. R-loops regulate both transcription

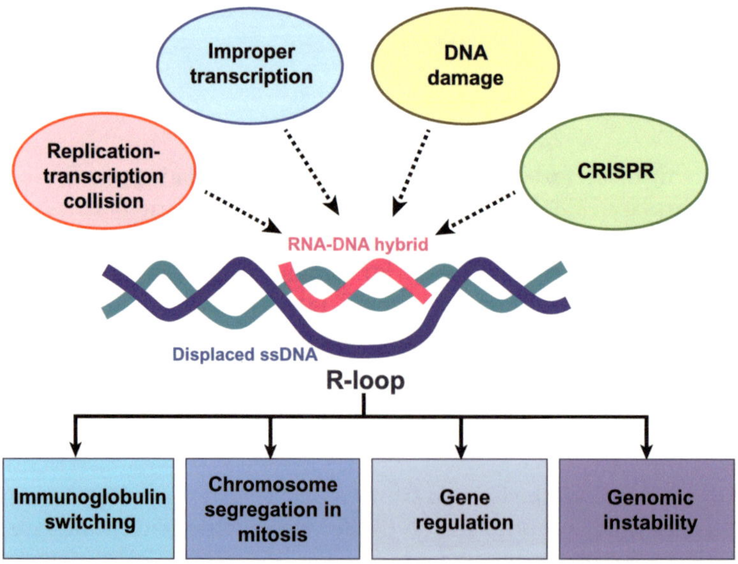


**Figure 1.**
*The causes and consequences of R-loop formation.*

initiation and termination [20]. Skourti-Stathaki *et al*. proposed a pause-dependent transcription termination mechanism mediated by R-loops and H3K9me2. R-loops are formed in transcription termination regions containing GC-rich sequences and facilitate antisense RNA transcription, inducing dsRNA for RNA interference factors, and they recruit G9a/GLP for the methylation of H3K9 along with HP1γ, which terminates transcription [21]. In addition to research by Skourti-Stathaki and colleagues, other studies suggest that R-loops can regulate transcription termination by RNA polymerase pausing [22, 23]. R-loops also occur at telomeres. Telomeric-repeat-containing RNAs (TERRAs) are noncoding RNAs transcribed from eukaryotic telomeres [24]. During telomerase-mediated telomere elongation in *rat1–1* mutant *Saccharomyces cerevisiae*, TERRAs form RNA–DNA hybrids and inhibit telomerase function [25]. In addition, the THO complex maintains yeast telomeres by suppressing R-loops generated by TERRAs [26].

### 1.3 R-loops, genomic instability, and human diseases

Despite the multiple roles of R-loops in normal cellular processes described above, they are also considered a form of DNA damage that can threaten genomic maintenance and integrity. In particular, the displaced ssDNA in an R-loop can increase genomic instability because it is a good endonuclease substrate [13, 27]. R-loops also induce replication stress. When the displaced ssDNA is broken, the replication fork stops at the R-loop. The RNA–DNA hybrid itself can block the progression of replication forks, and DNA polymerases may become trapped at R-loops [13, 28, 29]. Such replication stresses will activate DNA repair pathways, which might cause chromosome rearrangement [29].

Genomic instability that stems from R-loops may also contribute to some human diseases. Although there is no apparent evidence that R-loops are directly associated with disease, efforts to show causality between R-loops and disease have increased [30]. Some genetic disorders are caused by gene-specific repeats. R-loop formation is highly probable in tandem repeat sequences with high GC content and could change the repeat length. In particular, trinucleotide repeat expansion is a major cause of neurological and neuromuscular diseases, such as Huntington's disease and fragile X syndrome [31, 32]. It has been proposed that R-loops are associated with other neurological disorders, including amyotrophic lateral sclerosis, Aicardi-Goutières syndrome, and Prader-Willi syndrome [33–35]. R-loops also appear to be associated with cancer. *BRCA* genes, which are involved in DNA double-strand break repair via homologous recombination, are intimately associated with breast and ovarian cancer, and BRCA2 prevents R-loop accumulation [36, 37]. VIM is a member of the intermediate filament family and associated with different types of cancer. In colon cancer, the *VIM* promoter is hypermethylated and *VIM* expression is silenced. In normal cells, the gene is activated by R-loop formation in the promoter region, raising the possibility that transcription regulation by R-loops related to cancer development [18].

### 1.4 R-loop prevention and elimination

As described above, R-loops can cause genomic instability unless they are resolved, so they must be properly regulated. Several proteins are involved in R-loop prevention or elimination, such as RNase H, DNA TOP1, and Sen1 [38]. For example, RNase H directly removes R-loops by specifically degrading the RNA in RNA–DNA hybrid structures [39]. RNA helicases also resolve R-loops by unwinding RNA–DNA hybrid structures [40]. Because negative supercoiling promotes R-loop formation, topoisomerases play a key role in preventing R-loops [41]. In the case of replication fork stalling due to R-loops, FANCD2 recruits RNA processing enzymes such as hnRNP U and DDX47 to resolve R-loops at the stalled fork [42].

## 2. *In vivo* R-loop assays

Visualizing R-loop formation is important for understanding R-loop metabolism. Because R-loops basically consist of nucleic acids, distinguishing R-loop from ds- and ss-DNA or RNA using existing DNA staining or visualization methods is challenging. S9.6 is an antibody specific to an RNA–DNA hybrid, which was developed in 1986 and rapidly advanced R-loop-related research [43]. This antibody is commonly used to detect R-loops both *in vivo* and *in vitro* [44–47].

Currently, the most popular R-loop characterization technique is DNA–RNA immunoprecipitation sequencing (DRIP-seq), in which RNA–DNA hybrid strands are immunoprecipitated with S9.6 and then sequenced (**Figure 2**, [47, 48]). DRIP-seq was first adopted for profiling CpG island promoters, where R-loops are predominantly formed [16]. This method revealed that genes containing terminal GC-rich sequences form R-loops at their 3′-end, suggesting that R-loops contribute to efficient transcription termination [20]. The DRIP-seq technique has been further improved; S1 nuclease treatment prior to DRIP-seq can stabilize the DNA–RNA hybrid because S1 removes the displaced ssDNA, thus improving the resolution [49]. In conventional DRIP-seq, it is assumed that the content of R-loops or RNA–DNA hybrids does not vary depending on cell type and growth condition. For appropriate comparison, quantitative differential DNA–RNA immunoprecipitation sequence (qDRIP-seq) uses synthetic RNA–DNA hybrids as internal standards and facilitates comparison between different conditions with high resolution and sensitivity [50]. Although DRIP-seq is a very robust and well-characterized technique, it can only measure the ensemble average level of R-loops. However, single-molecule R-loop footprinting (SMRF-seq) can reveal the R-loop population via chemical reactivity of ssDNA at the single-molecule level. Malig *et al*. developed SMRF-sequencing based on non-denaturing bisulfite conversion [51, 52]. Sodium bisulfite treatment converts unpaired cytosines to uracils on ssDNA. In R-loops, only one strand is unpaired and exposed to bisulfite, whereas the complementary strand is protected by RNA. Thus, the PCR product of the displaced single strand in an R-loop has a converted sequence of cytosines to thymines, which is a footprint of the R-loop. PCR products are rapidly sequenced using a single-molecule real-time sequencing technique [53]. This approach enables characterization of the individual footprints of R-loops on long-range genomes at high resolution, even at the single-molecule level.

Fluorescently labeled S9.6 can be used as an R-loop probe in microscope imaging at the cellular level (**Figure 3a**). The brief procedure is following. K562 cells were pelleted after trypsinization for detaching cells from the plate. Supernatant was discarded to approximately 300 ul, and cell pellets were resuspended. 5 ml of 37°C pre-warmed 75 mM KCl solution was added in a drop-wise manner while the resuspended cells were slowly agitated. After the cells were incubated at 37°C for 14 min, five or six drops of fresh ice-cold fixative solution (3,1 methanol:acetic acid) were added to the cells, which were centrifuged again. Supernatant was discarded to approximately 300 ul, and cell pellets were resuspended. The cells were treated on ice with 5 ml of ice-cold fixative solution in a drop-wise manner. After washed once with

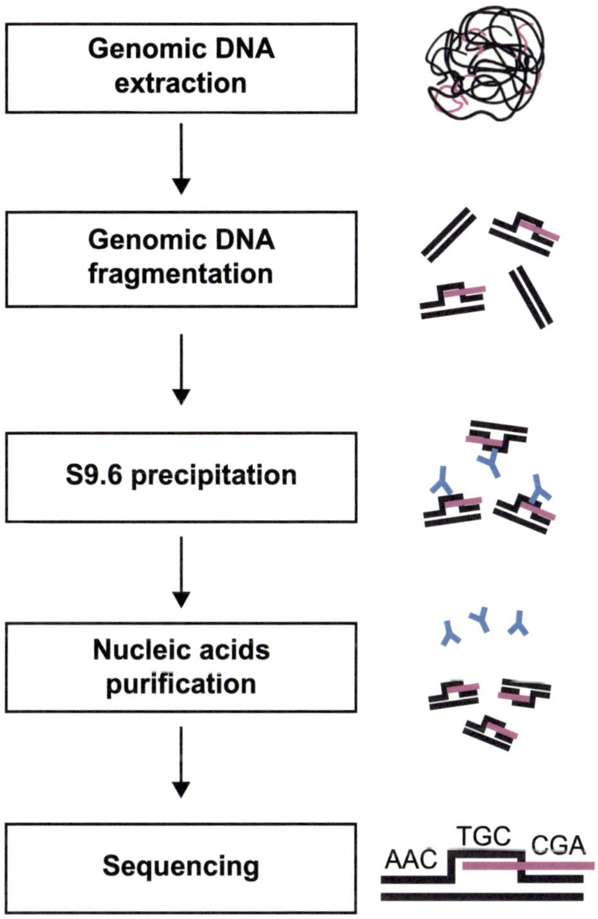


**Figure 2.**
*The flow chart of DRIP-seq. Step 1: whole-genomic DNA containing R-loops is extracted from cells. Step 2: extracted genomic DNA is fragmented by various types of restriction enzymes. Step 3: R-loops containing RNA–DNA hybrids are precipitated with S9.6 antibodies. Step 4: S9.6 antibodies are eliminated by proteinase K treatment, and R-loops are purified by phenol-chloroform extraction followed by ethanol precipitation. Step 5: precipitated R-loops are sequenced.*

fixative solution, the cells were spread onto a clean slide followed by 1 min incubation in 95°C steam for drying. The slide was immediately treated with blocking buffer (1x PBS, 5% BSA, 0.5% Triton X-100) and incubated at room temperature for 1 hr. The slide was successively treated with S9.6 antibody (1500) in blocking buffer at 4°C overnight. After residual S9.6 antibody was washed three times with washing buffer (1x PBS supplemented with 0.1% Triton X-100), the slide was treated with mouse AlexaFluor 594-conjugated secondary antibody at room temperature for 1 hr. The unbound secondary antibody was washed three times with washing buffer, and then the cells were stained with 4,6 diamidino-2-phenylindole (DAPI) and mounted using Vectashield (Vector Laboratories). Finally, the cells were imaged using a fluorescence microscope.

The use of immunofluorescence with S9.6 can allow visualization of the intracellular locations of RNA–DNA hybrids, even in mitochondria [54–56]. Furthermore, R-loop detection by S9.6 is ensured by RNase H1 overexpression (**Figure 3b**). R-loops

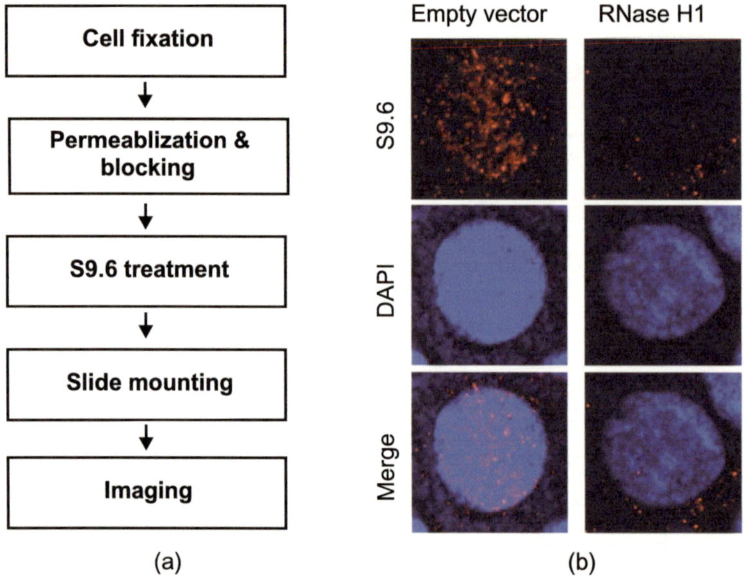


**Figure 3.**
*Flow chart of* in vivo *immunofluorescence imaging using S9.6 for R-loop visualization. (a) Flow chart of* in vivo *immunofluorescence imaging using S9.6 for R-loop visualization. (b) Cellular images of S9.6 labeled with fluorescent secondary antibodies. RNase H1 overexpression significantly reduces the S9.6 signal due to elimination of R-loops.*

can also be visualized via R-loop associated proteins with diverse modifications. Prendergast *et al*. subcloned the RNA binding domain (RBD) of RNase H1 fused to DsRed fluorescent protein to monitor intracellular R-loop dynamics [57]. Hodroj *et al*. enhanced green fluorescent protein (eGFP)-fused Ddx19 RNA helicase that specifically binds RNA–DNA hybrids. The fluorescent signal of eGFP-Ddx19 indicates R-loop formation inside cells. In addition, it does not form foci when RBD-mutated Ddx19 and phosphorylation site-mutated Ddx19 are used [58].

## 3. *In vitro* approaches

In addition to *in vivo* methods, several biochemical assays have been developed to study R-loops. Classically, R-loops are formed from transcription in a supercoiled plasmid by phage RNA polymerases (RNAPs) such as T3 or T7 [59]. Because R-loops have a three-strand structure, they have lower mobility than DNA duplexes in gel electrophoresis, so band shift or smearing occurs between supercoiled and relaxed plasmids during this procedure [53, 60]. Because RNase H digests ssRNA, dsRNA, and RNA–DNA hybrids, RNase H treatment eliminates the mobility shift of plasmids observed with gel electrophoresis [61, 62]. In addition, R-loops can be detected by isotope (e.g. $^{32}P$) or fluorescently labeled RNA, which is formed with isotope or fluorescently labeled ribonucleoside triphosphates during transcription. The labeled RNA is used to confirm if the plasmid mobility shift actually results from R-loops [53, 60]. In contrast, S9.6 is also used in electrophoresis mobility shift assay (EMSA) and Western blot with oligomers (**Figure 4a**). EMSA in **Figure 4a** was performed with fluorescently labeled R-loop and homoduplex DNA with synthesized oligomers (**Table 1**) following the previous protocol [55]. The oligomers were mixed for R-loop

*DOI: http://dx.doi.org/10.5772/intechopen.103176*

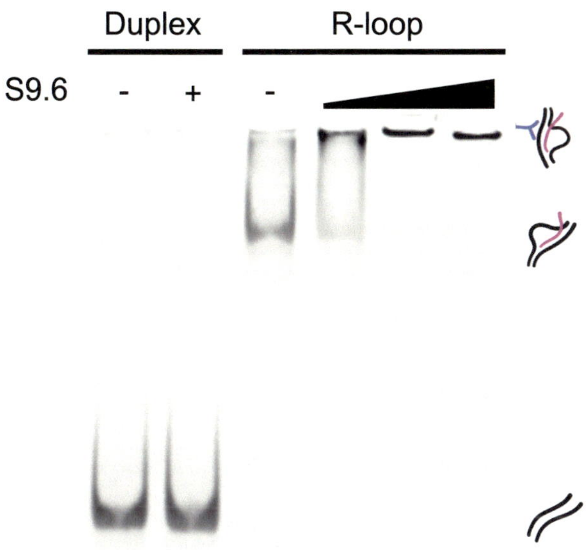


(a)

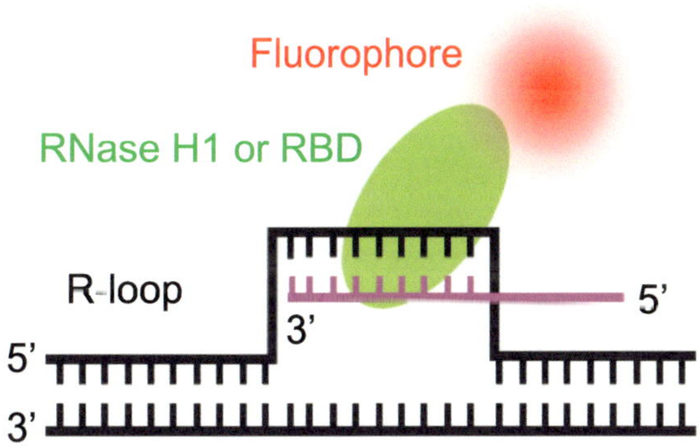


(b)

**Figure 4.**
*In vitro EMSA (electrophoresis mobility shift assay) for R-loop identification. (a) EMSA image for identifying R-loops in vitro. Fluorescently labeled oligomers were hybridized to form duplex DNA and R-loop structures. The R-loops shifted upward because of their lower mobility compared with same-length duplex DNA due to its molecular weight and the displaced ssDNA (lane 1 vs. lane 3). Because S9.6 specifically binds to R-loops, S9.6 treatment further super-shifts R-loops but not duplexes (lane 2 vs. lanes 4, 5, and 6). The black triangle represents the increasing concentration of S9.6 antibody (10, 30, and 50 nM). (b) Simplified diagram of R-loop visualization using fluorescently labeled catalytically-inactive RNase H1 or RBD. In addition to the S9.6 antibody, catalytically-inactive RNase H1 and RBD with fluorescence dye can also be used to visualize R-loops.*

and homoduplex DNA as shown in **Table 2**, and the mixture was heated at 95°C and then slowly cooled down to room temperature. 10 nM R-loop or duplex DNA was mixed to S9.6 (10, 30 and 50 nM) in reaction buffer (10 mM HEPES [pH 7.5], 1 mM DTT, and 5% glycerol). The reactant was incubated in the dark at room temperature for 20 min. The R-loop formation and the binding of S9.6 to R-loop were analyzed with 5% non-denaturing polyacrylamide gel electrophoresis and imaged by Typhoon

| Oligomer name | Sequences |
|---|---|
| DNA 1 | 5'-GCC GTC GCA TGA CGC TGC CGA ATT CTA CCA CGC GAT TCA TAC CTG TCG TGC CAG CTG CTT TGC CCA CCT GCA GGT TCA CCT CGT CCC TGG C-3′ |
| DNA 2 | 5′-[Cy3]-GCC AGG GAC GAG GTG AAC CTG CAG GTG GGC AAA GCA GCT GGC ACG ACA GGT ATG AAT CGC GTG GTA GAA TTC GGC AGC GTC ATG CGA CGG C-3' |
| DNA 3 | 5′-[Cy3]-GCC AGG GAC GAG GTG AAC CTG CAG GTG GGC GGC TAC TAC TTA GAT GTC ATC CGA GGC TTA TTG GTA GAA TTC GGC AGC GTC ATG C GA CGG C-3' |
| RNA 1 | 5′-[Cy5]-GCA GCU GGC ACG ACA GGU AUG AAU C-3' |

*[Cy3] and [Cy5] indicate the labeling of Cy3 and Cy5 fluorescent dyes, respectively.*

**Table 1.**
*List of oligomers for EMSA.*

| Substrate name | Mixture recipe (total 20 μl in reaction buffer) |
|---|---|
| **Homoduplex** | **DNA 1**: 6 μM, **DNA 2**: 5 μM |
| **R-Loop** | **DNA 1**: 6 μM, **DNA 3**: 5 μM, **RNA 1**: 5 μM |

**Table 2.**
*Hybridization of oligomers for EMSA.*

RGB (GE Healthcare) system. In EMSA using the oligomers, R-loops show band shift from dsDNA resulting from low mobility due to its triple-strand structure. Further band shift is observed following treatment with S9.6, which binds to RNA–DNA hybrids [50, 55]. In Western blotting, RNA–DNA hybrid interactors can be validated using S9.6 and target protein immunoprecipitation [45].

Purified RBD-DsRed specifically binds to RNA-containing structures, enabling its use as a probe of R-loops (**Figure 4b**). DNA fibers can be spread on a surface to distinguish the positions of R-loop regions with the purified RBD-DsRed. Various tags combining the RBD of RNase H1 have been used in microscopic fluorescent imaging, EMSA, and DRIP-seq to identify R-loops [63]. Crossely *et al.* designed and purified different types of RNase H1 that contains RBD and full amino acid sequences [50]. However, the RNase H1 used in their research has a D210N mutation that renders it catalytically inactive. Their construct successfully recognizes R-loops and RNA–DNA hybrids without degradation of RNA in EMSA. The GFP-labeled catalytic mutant RNase H1 thoroughly colocalized with R-loop-containing oligos within the cells.

Atomic force microscopy (AFM) scans a sample on a mica surface using a cantilever to yield a topographic image of the sample [64]. AFM has revealed diverse types of nucleic acids structures and DNA-protein complex formations [65–67]. AFM is also applied for visualizing R-loop formation. Carrasco-Salas *et al.* used AFM to observe three distinct structures derived from R-loops: blobs, spurs, and loops [68]. The specific R-loop structures depend on the sequence of non-template strand that is displaced in the R-loop, which suggests that non-template strand organization is an intrinsic characteristic of R-loops.

## 4. Single-molecule approaches for R-loop studies

Although R-loop formation, function, and fate have been extensively studied using biochemical assays and cell-based imaging as described above, those approaches still have limitations related to probing molecular details due to the ensemble average effect. Such hurdles can be overcome with single-molecule techniques that enable researchers to 1) observe individual molecules without an ensemble average effect, 2) mechanically manipulate biomolecules, and 3) directly observe biomolecular interactions [69]. Several single-molecule techniques have been utilized for R-loop studies. Lee *et al*. used protein-induced fluorescence enhancement (PIFE) to observe R-loop formation during T7 RNA polymerase transcription [70]. PIFE is a phenomenon in which a protein sometimes enhances the intensity of fluorescent dyes on other biomolecules [71]. PIFE assays exploit this intensity enhancement to measure the distance and interaction between non-tagged proteins and fluorescent dyes on target molecules, such as DNA. Fluorescent tagging of proteins is inefficient and may disturb protein activity; however, in PIFE assays there is no need to tag proteins [72]. The authors demonstrated that a G-quadruplex on the non-template strand stabilizes the R-loop, which enhances transcription elongation.

In addition to PIFE, single-molecule FRET (smFRET) has been widely used for probing the conformational dynamics of biomolecules (**Figure 5a**) [73, 74]. FRET requires two dyes (donor and acceptor) with spectral overlap for donor emission and acceptor absorption. In FRET, only the donor dye is excited, while the acceptor dye emits fluorescence through energy transfer when both dyes are in close proximity, as the energy transfer efficiency depends on the distance between them. R-loops are also studied using smFRET, during which the target DNA or RNA and RNA polymerases are fluorescently labeled with FRET donors and acceptors (**Figure 5a**, [75]). For smFRET experiments, one DNA oligomer with both FRET donor (Cy3) and acceptor (Cy5) and its complementary oligomer with biotin were hybridized. The hybridized DNA was anchored on the surface of a quartz slide coated with polyethylene glycol (PEG) via biotin-streptavidin interaction. Transcription was initiated by injecting 8 nM T7 RNA polymerases and 2 mM of rNTPs in imaging buffer (40 mM Tris–HCl [pH 8.0], 50 mM KCl, 5 mM NaOH, 20 mM $MgCl_2$, 1 mM DTT, 2 mM spermidine, 3 mM Trolox, 5 mM PCA, and 4 units/ml PCD). Total internal reflection fluorescence (TIRF) microscopy equipped with an electron-multiplying CCD camera was used for fluorescence imaging. Donor (Cy3) and acceptor (Cy5) dyes were excited by 532-nm and 633-nm lasers, respectively. smFRET experiments revealed that R-loop formation precedes and facilitates G-quadruplex formation, which is extremely stable even after R-loop resolution. Using smFRET, we can examine R-loop formation induced by dsDNA denaturation, collision between RNAP and obstacles such as protein roadblocks or DNA lesions, and G-quadruplex formation of displaced ssDNA during R-loop formation [70, 75].

In addition to R-loop formation, sensing R-loops is important for downstream processes, including R-loop resolution. In particular, how R-loop-binding proteins recognize R-loops in long genomic DNA is unclear. R-loop search mechanisms have been investigated with a novel single-molecule fluorescence imaging technique called DNA curtain (**Figure 5b**, [76, 77]). In this assay, DNA molecules are anchored on a lipid bilayer and aligned at nanometric diffusion barriers. Owing to the fluidity of the surface lipid bilayer, DNA molecules are unidirectionally stretched under hydrodynamic flow. DNA curtains can be used to identify sequence-dependent binding of proteins to DNA. Moreover, they allow us to visualize the movement of a protein along a single DNA molecule in real time. To study the search mechanism, an R-loop is inserted into a specific location of lambda phage DNA and fluorescently imaged with Cy5-labeled RNA in the R-loop. Then, the R-loop-binding protein is tagged with a fluorescent nanoparticle called a quantum dot (Qdot), which has a different

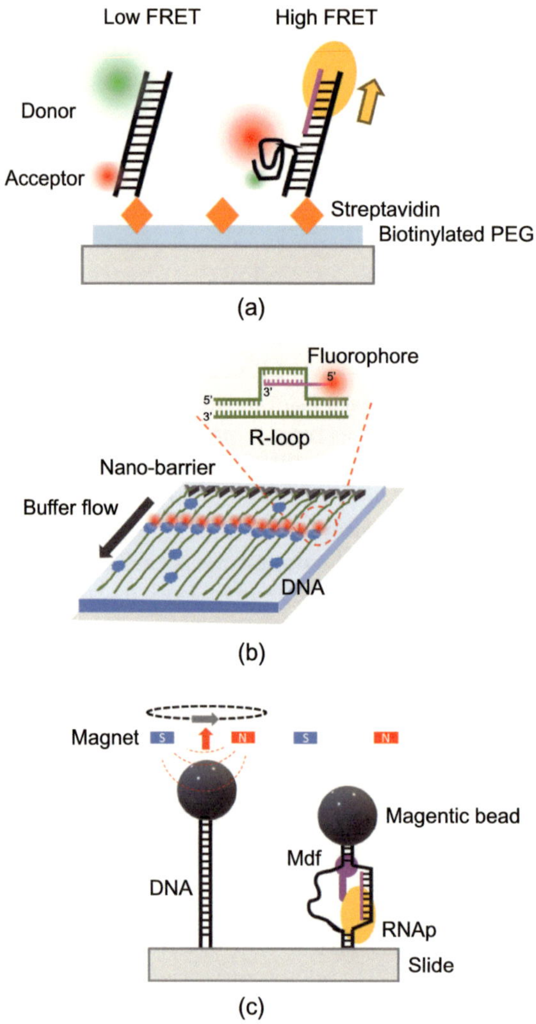


**Figure 5.**
*Single-molecule R-loop visualization techniques. (a) Schematic of single-molecule FRET. Hybridized oligomers are anchored on the biotinylated polyethylene glycol surface via biotin-streptavidin linkage. Donor (green) and acceptor (red) dyes in a duplex DNA display low FRET due to the long distance between the two dyes. However, when an R-loop is formed during the transcription by RNAP (yellow), the dissociated ssDNA emits high FRET. (b) Schematic of single-molecule DNA curtain. For DNA curtain assay, the slide surface with nanometric diffusion barriers was coated with a biotinylated lipid bilayer, which is made of DOPC (1,2-dioleoyl-sn-glycero-phosphocholine), 0.5% biotinylated-DPPE (1,2-dipalmitoyl-sn-glycero-3-phosphoethanolamine-N-(cap biotinyl)), and 8% mPEG 2000-DOPE (1,2-dioleoyl-sn-glycero-3-phosphoethanolamine-N-[methoxy(polyethylene glycol)-2000]). A Cy5-labeled R-loop was inserted into lambda phage DNA, which has biotin at one end. The lambda phage DNA was anchored on the lipid bilayer via biotin-streptavidin linkage in BSA buffer (40 mM Tris–HCl [pH 8.0], 50 mM NaCl, 2 mM $MgCl_2$, and 0.4% BSA). TonEBP with 3x FLAG was labeled with anti-FLAG conjugated quantum dot. Under hydrodynamic flow, DNA curtain was formed in reaction buffer (10 mM HEPES [pH 7.5] and 50 mM NaCl), and R-loops were imaged by Cy5 fluorescence under TIRF microscopy. Then quantum dot-labeled TonEBP was incubated with the lambda DNA, and its binding to the R-loop was imaged. DNA molecules containing an R-loop are unidirectionally stretched on the biotinylated lipid-coated slide (sky blue) and aligned at the chromium nano-barrier (gray) due to the fluidity of lipid bilayer. The interaction between TonEBP (blue) and R-loops (red), in which RNA is labeled with a fluorescent dye, can be visualized in real time. (c) Schematic of magnetic tweezers. The magnetic field exerts and measures both force and torque on the magnetic bead (black). The interaction between Mdf (violet) linked to both duplex DNA and RNAP (yellow) during R-loop formation can be measured based on the length change of DNA under a constant force using magnetic tweezers.*

emission wavelength from Cy5. Two-color imaging of both Cy5 and Qdot in the DNA curtain allows the R-loop search mechanism of the R-loop-binding protein. Kang *et al*. reported that tonicity enhancer-binding protein (TonEBP) plays important roles in R-loop sensing and recruitment of downstream proteins [55]. Using the DNA curtain approach, they revealed that TonEBP preferentially binds R-loops through both three-dimensional collision and one-dimensional diffusion. This dual-search mechanism facilitates rapid searches for R-loop throughout the long human genome. Furthermore, the quantitative analysis on one-dimensional diffusion shows that TonEBP diffuses along DNA by sliding rather than hopping. In EMSAs, TonEBP preferentially binds R-loops, D-loops, and bubble DNA structures over duplex DNA. The substances for which TonEBP has a high affinity all contain displaced ssDNA structures. These results indicate that TonEBP preferentially binds displaced ssDNA, thus recognizing R-loops on duplex DNA.

Magnetic tweezers assay can measure both the tension and topological features of a single supercoiled DNA. In this approach, a linear DNA molecule is torsionally constrained by tethering the DNA ends to the slide surface and a magnetic bead that is rotated to induce DNA supercoiling (**Figure 5c**). Portmen *et al*. used magnetic tweezers to elucidate the R-loop formation mechanism by the transcription-coupled repair factor Mfd during transcription based on topologically changing the DNA [78]. For the magnetic tweezers assay, 4.6 kbp long DNA containing a promoter site was ligated with biotinylated handle at one end and digoxigenin handle at the other hand. Digoxigenin end was anchored on an anti-digoxigenin-coated glass coverslip, and biotinylated end was attached to a 1 μm diameter superparamagnetic bead. Transcription reaction was done in reaction buffer (40 mM HEPES [pH 8.0], 100 mM KCl, 8 mM $MgCl_2$, 0.5 mg/ml BSA, and 1 mM DTT) supplemented with 300 pM RNA polymerases, 500 nM Mfd, 50 nM GreB, 1 mM rATP, 100 μM rCTP, 100 μM rUTP, and 100 μM rGTP. They found that R-loop formation was mediated by the Mfd-RNAP complex, which compacted and supercoiled the template DNA during transcription. Mfd simultaneously binds both RNAP and DNA and results in tripartite supercoiled domains. The negative supercoiling in the tripartite domains serves as a substrate for R-loop formation.

With advances in single-molecule imaging technology, we can investigate R-loops and related factors that cannot be observed in traditional ensemble assays. The convergence of single-molecule techniques and R-loop research will pave the way to more thorough investigation of R-loops with higher spatiotemporal resolution.

## 5. Conclusions

R-loops are involved in various cellular activities but can threaten genomic stability. Detecting these structures is important for understanding their metabolism and underlying mechanism. This chapter described the formation, roles, and regulation of R-loops and related diseases and explored *in vivo* and *in vitro* methods for R-loop detection and visualization, including single-molecule techniques. The most classical methods for R-loop are based on the S9.6 antibody. However, novel techniques that do not require this antibody have been developed. In particular, single-molecule R-loop imaging techniques have accelerated research. We expect that more advanced techniques for R-loops with high sensitivity and resolution will be developed in the future.

## Acknowledgements

This work was supported by the National Research Foundation (grant number: NRF-2020R1A2B5B01001792) and the Institute for Basic Science (IBS-R022-D1).

## Conflict of interest

The authors declare no conflict of interest.

## Author details

Na Young Cheon[1], Subin Kim[1] and Ja Yil Lee[1,2]*

1 Department of Biological Sciences, Ulsan National Institute of Science and Technology, Ulsan, Republic of Korea

2 Center for Genomic Integrity, Institute for Basic Science, Ulsan, Republic of Korea

*Address all correspondence to: biojayil@unist.ac.kr

## References

[1] Thomas M, White RL, Davis RW. Hybridization of RNA to double-stranded DNA: Formation of R-loops. Proceedings of the National Academy of Sciences of the United States of America. 1976;**73**(7):2294-2298. DOI: 10.1073/pnas.73.7.2294

[2] Roberts RW, Crothers DM. Stability and properties of double and triple helices: Dramatic effects of RNA or DNA backbone composition. Science. 1992;**258**(5087):1463-1466. DOI: 10.1126/science.1279808

[3] Wang AH, Fujii S, van Boom JH, van der Marel GA, van Boeckel SA, Rich A. Molecular structure of r(GCG)d(TATACGC): A DNA--RNA hybrid helix joined to double helical DNA. Nature. 1982;**299**(5884):601-604. DOI: 10.1038/299601a0

[4] Ehara H, Sekine SI. Architecture of the RNA polymerase II elongation complex: New insights into Spt4/5 and Elf1. Transcription. 2018;**9**(5):286-291. DOI: 10.1080/21541264.2018.1454817

[5] Zhang B, Luo D, Li Y, Perculija V, Chen J, Lin J, et al. Mechanistic insights into the R-loop formation and cleavage in CRISPR-Cas12i1. Nature Communications. 2021;**12**(1):3476. DOI: 10.1038/s41467-021-23876-5

[6] Martinez-Rucobo FW, Cramer P. Structural basis of transcription elongation. Biochimica et Biophysica Acta. 2013;**1829**(1):9-19. DOI: 10.1016/j.bbagrm.2012.09.002

[7] Yin YW, Steitz TA. The structural mechanism of translocation and helicase activity in T7 RNA polymerase. Cell. 2004;**116**(3):393-404. DOI: 10.1016/s0092-8674(04)00120-5

[8] Roy D, Zhang Z, Lu Z, Hsieh CL, Lieber MR. Competition between the RNA transcript and the nontemplate DNA strand during R-loop formation in vitro: A nick can serve as a strong R-loop initiation site. Molecular and Cellular Biology. 2010;**30**(1):146-159. DOI: 10.1128/MCB.00897-09

[9] Liu LF, Wang JC. Supercoiling of the DNA template during transcription. Proceedings of the National Academy of Sciences of the United States of America. 1987;**84**(20):7024-7027. DOI: 10.1073/pnas.84.20.7024

[10] Wahba L, Amon JD, Koshland D, Vuica-Ross M. RNase H and multiple RNA biogenesis factors cooperate to prevent RNA:DNA hybrids from generating genome instability. Molecular Cell. 2011;**44**(6):978-988. DOI: 10.1016/j.molcel.2011.10.017

[11] Dominguez-Sanchez MS, Barroso S, Gomez-Gonzalez B, Luna R, Aguilera A. Genome instability and transcription elongation impairment in human cells depleted of THO/TREX. PLoS Genetics. 2011;**7**(12):e1002386. DOI: 10.1371/journal.pgen.1002386

[12] Gomez-Gonzalez B, Garcia-Rubio M, Bermejo R, Gaillard H, Shirahige K, Marin A, et al. Genome-wide function of THO/TREX in active genes prevents R-loop-dependent replication obstacles. The EMBO Journal. 2011;**30**(15):3106-3119. DOI: 10.1038/emboj.2011.206

[13] Huertas P, Aguilera A. Cotranscriptionally formed DNA:RNA hybrids mediate transcription elongation impairment and transcription-associated recombination. Molecular Cell. 2003;**12**(3):711-721. DOI: 10.1016/j.molcel.2003.08.010

[14] Roy D, Yu K, Lieber MR. Mechanism of R-loop formation at immunoglobulin class switch sequences. Molecular and Cellular Biology. 2008;**28**(1):50-60. DOI: 10.1128/MCB.01251-07

[15] Deaton AM, Bird A. CpG islands and the regulation of transcription. Genes & Development. 2011;**25**(10):1010-1022. DOI: 10.1101/gad.2037511

[16] Ginno PA, Lott PL, Christensen HC, Korf I, Chedin F. R-loop formation is a distinctive characteristic of unmethylated human CpG island promoters. Molecular Cell. 2012;**45**(6):814-825. DOI: 10.1016/j.molcel.2012.01.017

[17] Sun Q, Csorba T, Skourti-Stathaki K, Proudfoot NJ, Dean C. R-loop stabilization represses antisense transcription at the Arabidopsis FLC locus. Science. 2013;**340**(6132):619-621. DOI: 10.1126/science.1234848

[18] Boque-Sastre R, Soler M, Oliveira-Mateos C, Portela A, Moutinho C, Sayols S, et al. Head-to-head antisense transcription and R-loop formation promotes transcriptional activation. Proceedings of the National Academy of Sciences of the United States of America. 2015;**112**(18):5785-5790. DOI: 10.1073/pnas.1421197112

[19] Castellano-Pozo M, Santos-Pereira JM, Rondon AG, Barroso S, Andujar E, Perez-Alegre M, et al. R loops are linked to histone H3 S10 phosphorylation and chromatin condensation. Molecular Cell. 2013;**52**(4):583-590. DOI: 10.1016/j.molcel.2013.10.006

[20] Ginno PA, Lim YW, Lott PL, Korf I, Chedin F. GC skew at the 5′ and 3′ ends of human genes links R-loop formation to epigenetic regulation and transcription termination. Genome Research. 2013;**23**(10):1590-1600. DOI: 10.1101/gr.158436.113

[21] Skourti-Stathaki K, Kamieniarz-Gdula K, Proudfoot NJ. R-loops induce repressive chromatin marks over mammalian gene terminators. Nature. 2014;**516**(7531):436-439. DOI: 10.1038/nature13787

[22] Grzechnik P, Gdula MR, Proudfoot NJ. Pcf11 orchestrates transcription termination pathways in yeast. Genes & Development. 2015;**29**(8):849-861. DOI: 10.1101/gad.251470.114

[23] Skourti-Stathaki K, Proudfoot NJ, Gromak N. Human senataxin resolves RNA/DNA hybrids formed at transcriptional pause sites to promote Xrn2-dependent termination. Molecular Cell. 2011;**42**(6):794-805. DOI: 10.1016/j.molcel.2011.04.026

[24] Balk B, Maicher A, Dees M, Klermund J, Luke-Glaser S, Bender K, et al. Telomeric RNA-DNA hybrids affect telomere-length dynamics and senescence. Nature Structural & Molecular Biology. 2013;**20**(10):1199-1205. DOI: 10.1038/nsmb.2662

[25] Luke B, Panza A, Redon S, Iglesias N, Li Z, Lingner J. The Rat1p 5′ to 3′ exonuclease degrades telomeric repeat-containing RNA and promotes telomere elongation in Saccharomyces cerevisiae. Molecular Cell. 2008;**32**(4):465-477. DOI: 10.1016/j.molcel.2008.10.019

[26] Pfeiffer V, Crittin J, Grolimund L, Lingner J. The THO complex component Thp2 counteracts telomeric R-loops and telomere shortening. The EMBO Journal. 2013;**32**(21):2861-2871. DOI: 10.1038/emboj.2013.217

[27] Basu U, Meng FL, Keim C, Grinstein V, Pefanis E, Eccleston J, et al.

The RNA exosome targets the AID cytidine deaminase to both strands of transcribed duplex DNA substrates. Cell. 2011;**144**(3):353-363. DOI: 10.1016/j.cell.2011.01.001

[28] Li X, Manley JL. Cotranscriptional processes and their influence on genome stability. Genes & Development. 2006;**20**(14):1838-1847. DOI: 10.1101/gad.1438306

[29] Aguilera A, Garcia-Muse T. R loops: From transcription byproducts to threats to genome stability. Molecular Cell. 2012;**46**(2):115-124. DOI: 10.1016/j.molcel.2012.04.009

[30] Richard P, Manley JL. R loops and links to human disease. Journal of Molecular Biology. 2017;**429**(21):3168-3180. DOI: 10.1016/j.jmb.2016.08.031

[31] Reddy K, Tam M, Bowater RP, Barber M, Tomlinson M, Nichol Edamura K, et al. Determinants of R-loop formation at convergent bidirectionally transcribed trinucleotide repeats. Nucleic Acids Research. 2011;**39**(5):1749-1762. DOI: 10.1093/nar/gkq935

[32] Reddy K, Schmidt MH, Geist JM, Thakkar NP, Panigrahi GB, Wang YH, et al. Processing of double-R-loops in (CAG).(CTG) and C9orf72 (GGGGCC).(GGCCCC) repeats causes instability. Nucleic Acids Research. 2014;**42**(16):10473-10487. DOI: 10.1093/nar/gku658

[33] Lim YW, Sanz LA, Xu X, Hartono SR, Chedin F. Genome-wide DNA hypomethylation and RNA:DNA hybrid accumulation in Aicardi-Goutieres syndrome. eLife. 2015;**4**:e08007. DOI: 10.7554/eLife.08007

[34] Powell WT, Coulson RL, Gonzales ML, Crary FK, Wong SS, Adams S, et al. R-loop formation at Snord116 mediates topotecan inhibition of Ube3a-antisense and allele-specific chromatin decondensation. Proceedings of the National Academy of Sciences of the United States of America. 2013;**110**(34):13938-13943. DOI: 10.1073/pnas.1305426110

[35] Salvi JS, Mekhail K. R-loops highlight the nucleus in ALS. Nucleus. 2015;**6**(1):23-29. DOI: 10.1080/19491034.2015.1004952

[36] Bhatia V, Barroso SI, Garcia-Rubio ML, Tumini E, Herrera-Moyano E, Aguilera A. BRCA2 prevents R-loop accumulation and associates with TREX-2 mRNA export factor PCID2. Nature. 2014;**511**(7509):362-365. DOI: 10.1038/nature13374

[37] Hatchi E, Skourti-Stathaki K, Ventz S, Pinello L, Yen A, Kamieniarz-Gdula K, et al. BRCA1 recruitment to transcriptional pause sites is required for R-loop-driven DNA damage repair. Molecular Cell. 2015;**57**(4):636-647. DOI: 10.1016/j.molcel.2015.01.011

[38] Costantino L, Koshland D. Genome-wide map of R-loop-induced damage reveals how a subset of R-loops contributes to genomic instability. Molecular Cell. 2018;**71**(4):487-97 e3. DOI: 10.1016/j.molcel.2018.06.037

[39] Cerritelli SM, Crouch RJ. Ribonuclease H: The enzymes in eukaryotes. The FEBS Journal. 2009;**276**(6):1494-1505. DOI: 10.1111/j.1742-4658.2009.06908.x

[40] Hamperl S, Cimprich KA. The contribution of co-transcriptional RNA:DNA hybrid structures to DNA damage and genome instability. DNA Repair (Amst). 2014;**19**:84-94. DOI: 10.1016/j.dnarep.2014.03.023

[41] Drolet M, Bi X, Liu LF. Hypernegative supercoiling of the DNA template during transcription elongation in vitro. The Journal of Biological Chemistry. 1994;**269**(3):2068-2074

[42] Okamoto Y, Abe M, Itaya A, Tomida J, Ishiai M, Takaori-Kondo A, et al. FANCD2 protects genome stability by recruiting RNA processing enzymes to resolve R-loops during mild replication stress. The FEBS Journal. 2019;**286**(1):139-150. DOI: 10.1111/febs.14700

[43] Boguslawski SJ, Smith DE, Michalak MA, Mickelson KE, Yehle CO, Patterson WL, et al. Characterization of monoclonal antibody to DNA.RNA and its application to immunodetection of hybrids. Journal of Immunological Methods. 1986;**89**(1):123-130. DOI: 10.1016/0022-1759(86)90040-2

[44] Kotsantis P, Silva LM, Irmscher S, Jones RM, Folkes L, Gromak N, et al. Increased global transcription activity as a mechanism of replication stress in cancer. Nature Communications. 2016;**7**:13087. DOI: 10.1038/ncomms13087

[45] Cristini A, Groh M, Kristiansen MS, Gromak N. RNA/DNA hybrid Interactome identifies DXH9 as a molecular player in transcriptional termination and R-loop-associated DNA damage. Cell Reports. 2018;**23**(6):1891-1905. DOI: 10.1016/j.celrep.2018.04.025

[46] Arab K, Karaulanov E, Musheev M, Trnka P, Schafer A, Grummt I, et al. GADD45A binds R-loops and recruits TET1 to CpG island promoters. Nature Genetics. 2019;**51**(2):217-223. DOI: 10.1038/s41588-018-0306-6

[47] Sanz LA, Castillo-Guzman D, Chedin F. Mapping R-loops and RNA:DNA hybrids with S9.6-based immunoprecipitation methods. Journal of Visualized Experiments. 2021;**174**. DOI: 10.3791/62455

[48] Sanz LA, Chedin F. High-resolution, strand-specific R-loop mapping via S9.6-based DNA-RNA immunoprecipitation and high-throughput sequencing. Nature Protocols. 2019;**14**(6):1734-1755. DOI: 10.1038/s41596-019-0159-1

[49] Wahba L, Costantino L, Tan FJ, Zimmer A, Koshland D. S1-DRIP-seq identifies high expression and polyA tracts as major contributors to R-loop formation. Genes & Development. 2016;**30**(11):1327-1338. DOI: 10.1101/gad.280834.116

[50] Crossley MP, Bocek MJ, Hamperl S, Swigut T, Cimprich KA. qDRIP: A method to quantitatively assess RNA-DNA hybrid formation genome-wide. Nucleic Acids Research. 2020;**48**(14):e84. DOI: 10.1093/nar/gkaa500

[51] Malig M, Hartono SR, Giafaglione JM, Sanz LA, Chedin F. Ultra-deep coverage single-molecule R-loop Footprinting reveals principles of R-loop formation. Journal of Molecular Biology. 2020;**432**(7):2271-2288. DOI: 10.1016/j.jmb.2020.02.014

[52] Malig M, Chedin F. Characterization of R-loop structures using single-molecule R-loop Footprinting and sequencing. Methods in Molecular Biology. 2020;**2161**:209-228. DOI: 10.1007/978-1-0716-0680-3_15

[53] Yu K, Chedin F, Hsieh CL, Wilson TE, Lieber MR. R-loops at immunoglobulin class switch regions in the chromosomes of stimulated B cells. Nature Immunology. 2003;**4**(5):442-451. DOI: 10.1038/ni919

[54] Silva S, Camino LP, Aguilera A. Human mitochondrial degradosome

prevents harmful mitochondrial R loops and mitochondrial genome instability. Proceedings of the National Academy of Sciences of the United States of America. 2018;**115**(43):11024-11029. DOI: 10.1073/pnas.1807258115

[55] Kang HJ, Cheon NY, Park H, Jeong GW, Ye BJ, Yoo EJ, et al. TonEBP recognizes R-loops and initiates m6A RNA methylation for R-loop resolution. Nucleic Acids Research. 2021;**49**(1):269-284. DOI: 10.1093/nar/gkaa1162

[56] Sollier J, Stork CT, Garcia-Rubio ML, Paulsen RD, Aguilera A, Cimprich KA. Transcription-coupled nucleotide excision repair factors promote R-loop-induced genome instability. Molecular Cell. 2014;**56**(6):777-785. DOI: 10.1016/j.molcel.2014.10.020

[57] Prendergast L, McClurg UL, Hristova R, Berlinguer-Palmini R, Greener S, Veitch K, et al. Resolution of R-loops by INO80 promotes DNA replication and maintains cancer cell proliferation and viability. Nature Communications. 2020;**11**(1):4534. DOI: 10.1038/s41467-020-18306-x

[58] Hodroj D, Recolin B, Serhal K, Martinez S, Tsanov N, Abou Merhi R, et al. An ATR-dependent function for the Ddx19 RNA helicase in nuclear R-loop metabolism. The EMBO Journal. 2017;**36**(9):1182-1198. DOI: 10.15252/embj.201695131

[59] Zaychikov E, Denissova L, Heumann H. Translocation of the Escherichia coli transcription complex observed in the registers 11 to 20: "jumping" of RNA polymerase and asymmetric expansion and contraction of the "transcription bubble". Proceedings of the National Academy of Sciences of the United States of America. 1995;**92**(5):1739-1743. DOI: 10.1073/pnas.92.5.1739

[60] Daniels GA, Lieber MR. RNA:DNA complex formation upon transcription of immunoglobulin switch regions: Implications for the mechanism and regulation of class switch recombination. Nucleic Acids Research. 1995;**23**(24):5006-5011. DOI: 10.1093/nar/23.24.5006

[61] Allison DF, Wang GG. R-loops: Formation, function, and relevance to cell stress. Cell Stress. 2019;**3**(2):38-46. DOI: 10.15698/cst2019.02.175

[62] Chon H, Vassilev A, DePamphilis ML, Zhao Y, Zhang J, Burgers PM, et al. Contributions of the two accessory subunits, RNASEH2B and RNASEH2C, to the activity and properties of the human RNase H2 complex. Nucleic Acids Research. 2009;**37**(1):96-110. DOI: 10.1093/nar/gkn913

[63] Wang K, Wang H, Li C, Yin Z, Xiao R, Li Q, et al. Genomic profiling of native R loops with a DNA-RNA hybrid recognition sensor. Science Advances. 2021;**7**(8):eabe3516. DOI: 10.1126/sciadv.abe3516

[64] Ohnesorge F, Binnig G. True atomic resolution by atomic force microscopy through repulsive and attractive forces. Science. 1993;**260**(5113):1451-1456. DOI: 10.1126/science.260.5113.1451

[65] Lyubchenko YL, Gall AA, Shlyakhtenko LS. Visualization of DNA and protein-DNA complexes with atomic force microscopy. Methods in Molecular Biology. 2014;**1117**:367-384. DOI: 10.1007/978-1-62703-776-1_17

[66] Lyubchenko YL, Shlyakhtenko LS. Imaging of DNA and protein-DNA complexes with atomic force microscopy. Critical Reviews in Eukaryotic Gene Expression. 2016;**26**(1):63-96. DOI: 10.1615/CritRevEukaryotGeneExpr.v26.i1.70

[67] Bustamante C, Vesenka J, Tang CL, Rees W, Guthold M, Keller R. Circular DNA molecules imaged in air by scanning force microscopy. Biochemistry. 1992;**31**(1):22-26. DOI: 10.1021/bi00116a005

[68] Carrasco-Salas Y, Malapert A, Sulthana S, Molcrette B, Chazot-Franguiadakis L, Bernard P, et al. The extruded non-template strand determines the architecture of R-loops. Nucleic Acids Research. 2019;**47**(13):6783-6795. DOI: 10.1093/nar/gkz341

[69] Kim SO, Jackman JA, Mochizuki M, Yoon BK, Hayashi T, Cho NJ. Correlating single-molecule and ensemble-average measurements of peptide adsorption onto different inorganic materials. Physical Chemistry Chemical Physics. 2016;**18**(21):14454-14459. DOI: 10.1039/c6cp01168c

[70] Lee CY, McNerney C, Ma K, Zhao W, Wang A, Myong S. R-loop induced G-quadruplex in non-template promotes transcription by successive R-loop formation. Nature Communications. 2020;**11**(1):3392. DOI: 10.1038/s41467-020-17176-7

[71] Hwang H, Kim H, Myong S. Protein induced fluorescence enhancement as a single molecule assay with short distance sensitivity. Proceedings of the National Academy of Sciences of the United States of America. 2011;**108**(18):7414-7418. DOI: 10.1073/pnas.1017672108

[72] Hwang H, Myong S. Protein induced fluorescence enhancement (PIFE) for probing protein-nucleic acid interactions. Chemical Society Reviews. 2014;**43**(4):1221-1229. DOI: 10.1039/c3cs60201j

[73] Ha T. Single-molecule fluorescence resonance energy transfer. Methods. 2001;**25**(1):78-86. DOI: 10.1006/meth.2001.1217

[74] Roy R, Hohng S, Ha T. A practical guide to single-molecule FRET. Nature Methods. 2008;**5**(6):507-516. DOI: 10.1038/nmeth.1208

[75] Lim G, Hohng S. Single-molecule fluorescence studies on cotranscriptional G-quadruplex formation coupled with R-loop formation. Nucleic Acids Research. 2020;**48**(16):9195-9203. DOI: 10.1093/nar/gkaa695

[76] Collins BE, Ye LF, Duzdevich D, Greene EC. DNA curtains: Novel tools for imaging protein-nucleic acid interactions at the single-molecule level. Methods in Cell Biology. 2014;**123**:217-234. DOI: 10.1016/B978-0-12-420138-5.00012-4

[77] Greene EC, Wind S, Fazio T, Gorman J, Visnapuu ML. DNA curtains for high-throughput single-molecule optical imaging. Methods in Enzymology. 2010;**472**:293-315. DOI: 10.1016/S0076-6879(10)72006-1

[78] Portman JR, Brouwer GM, Bollins J, Savery NJ, Strick TR. Cotranscriptional R-loop formation by Mfd involves topological partitioning of DNA. Proceedings of the National Academy of Sciences of the United States of America. 2021;**118**(15):e2019630118. DOI: 10.1073/pnas.2019630118

Chapter 4

# lncRNAs: Role in Regulation of Gene Expression

*Pranjal Kumar and Nikita Bhandari*

## Abstract

The long non-coding RNAs (lncRNAs) are a subclass of ncRNA which is more than 200 nucleotides long and processed similar to mRNA by RNA polymerase II with very few differences between them. In the last two decades, it has become a hot topic of research as it has been found differentially expressed in disease versus normal conditions including cancers. They regulate many biological functions including regulation of gene expression and epigenetic control. lncRNAs can control gene expression at the transcriptional level, and post-transcriptional level. Also, they can play a structural role to function as scaffolds for protein complexes. They interact with DNA, RNA, and proteins. They have been shown to possess competitive binding sites for miRNAs, which makes them a master regulator of gene expression by masking miRNAs and altering many biological functions. They are found to be associated with many cellular functions including cell proliferation, migration, and invasion. The lncRNAs can be utilized as biomarkers and can be targeted for personalized therapy.

**Keywords:** ncRNA, lncRNA, miRNA sponge, gene expression, biomarker, targeted therapy, epigenetic regulation, etc.

## 1. Introduction

Ribonucleic acid (RNA) is a biological macromolecule, which serves as genetic material as well as carries the information from deoxyribonucleic acid (DNA). RNAs are of multiple types and sub-categorized as per the functions they carry out, such as messenger RNA (mRNA), transfer RNA (tRNA), ribosomal RNA (rRNA), etc. Among these, mRNA is called coding RNA as it gets translated into protein while others are considered non-coding RNAs (ncRNAs) with no coding potential. The ncRNAs have been studied since the 1950s, which were mostly limited to tRNA and rRNA. Later, microRNAs (miRNAs) were discovered and to date, it is the most studied ncRNAs. ncRNAs were further subdivided as per their sizes. The term lncRNA stands for long non-coding RNA. It has been sub-categorized as lncRNA having more than 200 nucleotides [1, 2].

The human genome consists of more than 100,000 lncRNA genes [3, 4]. They are mainly transcribed by RNA polymerase II which leads to the 5′-capping i.e. (addition of 7-methyl guanosine at the 5′-end), and polyadenylation at the 3′-end [5]. The processing of lncRNAs transcription is similar to that of mRNAs. Earlier it was thought to be junk, but recent advancements in the understanding of ncRNAs have found it as regulatory biomolecules. Although the total number of lncRNAs is always debatable with the number of functionally characterized lncRNAs. With the recent findings, it is clear that they are part of many biological processes and their regulation.

There are an increasing number of lncRNAs that are functionally characterized, but still, there is a need to have more pieces of evidence to support the recent findings. These lncRNAs are associated with many cellular functions including gene expression and regulation. lncRNAs have been shown to control many gene expressions, some of them control only neighboring gene expressions while some function at a distant position. They have structural and functional roles in the regulation of gene expression. Recent studies over the last decade show that they are part of the regulatory roles in embryonic development [6] as well as in human diseases including cancers [7, 8], heart diseases [9], etc. The lncRNAs functions have been found to be associated at the transcriptional level, translational level, and chromatin levels [10]. In this chapter, we are going to review lncRNA biology, from their biogenesis to functions mainly in gene expression and regulation.

## 2. Genomic organization of lncRNAs

There are great numbers of noncoding regions (about 98–99% of the genome sequences in human) distributed between the coding region [11, 12]. The lncRNAs' sequences are present throughout in the genome and can be studied as different types depending upon their location in the genome. The lncRNAs can be broadly categorized in five types based on their genomic location [6]. They are stand-alone (also called as long intergenic non-coding RNA), natural antisense lncRNAs (tend to be highly enriched near promoter or terminator regions), pseudogenes (extra copy of existing genes, which are no more capable of coding), long intronic lncRNAs (synthesized from the intronic region of already annotated genes) and divergent lncRNAs (close to the transcription start site, promoter or enhancer regions). These all types of transcripts are presented in **Figure 1**. They are classified as different types based on their genomic positions but it is not on the basis of their function which simply help in the organizing the diverse species of lncRNAs [6].

The lncRNAs are synthesized from distinct genomic location, therefore, they have been named accordingly. The lncRNAs which come from intergenic regions are called as intergenic lncRNAs, also known as stand-alone lncRNAs. Similarly, the lncRNAs which are synthesized from intronic regions of a protein coding genes are called as intronic lncRNAs. All other types are also named as per their genomic location [6].

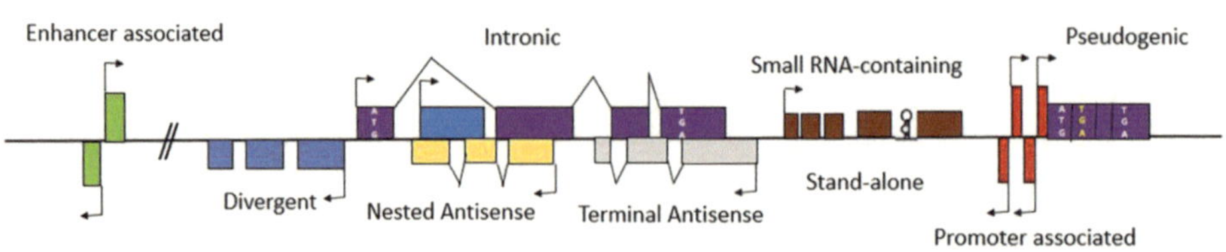


**Figure 1.**
*Genomic organization of lncRNAs. lncRNAs are named as per their relative location corresponding to already annotated protein coding genes. Arrow denotes the direction of lncRNA gene sequence at their Transcription Start Site. Protein Coding Genes (exons) are shown in purple boxes, various types of lncRNAs are shown in different colors. In case of pseudogenic lncRNAs, it is premature termination of transcription (yellow color TGA).*

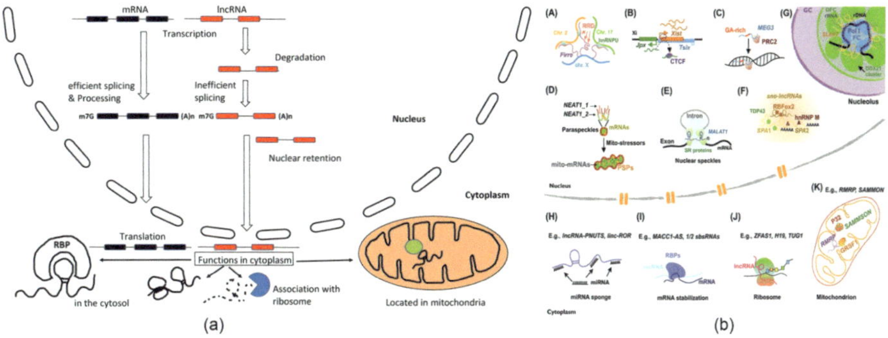


**Figure 2.**
*Biogenesis and Localization of IncRNAs. (a) Diagrammatic representation of IncRNA and mRNA transcription and the localization of IncRNAs in different cellular compartments for their function. (b) Some well studied IncRNAs at their respective cellular compartments (Chun-Jie Guo et al. 2020).*

## 3. Biogenesis and localization of lncRNAs

The lncRNAs are synthesized similarly to mRNA i.e. they are transcribed by RNA polymerase II, capped at 5′-end (m7G), poly-A tail at the 3′-end (polyadenylation), and spliced to remove introns. **Figure 2a** illustrates the comparison between the biogenesis of mRNAs and lncRNAs. **Figure 2b** (adapted from Chun-Jie Guo et al. 2020) [13] tells that a greater number of lncRNAs are found to be localized in the nucleus. The basic difference between the lncRNAs and mRNAs includes the sequence conservation and the number of exons. The lncRNAs are comparatively less conserved than mRNAs and they are composed of fewer exons [13–15].

The expression of lncRNAs is controlled by histone modification at their promoter regions [16, 17]. Also, the phosphorylation status of RNA polymerase II defines the expression of lncRNAs. They are also transcribed by dysregulated RNA polymerase II, which leads to the accumulation of some faulty lncRNAs on chromatin which are soon degraded by ribonuclease complex known as RNA exosomes [18]. lncRNAs are found to be localized in the nucleus as well as cytoplasm. One of the reasons for their nuclear retention is somewhat associated with weak splicing signals i.e. the length of the segment between the branch point and 3′ splice site is comparatively longer than the length in mRNA [17, 19, 20]. Other factors, including splicing inhibitors [13] and alternative poly-A signals, can also regulate the localization of some lncRNAs. Some of the lncRNAs, which localize to the cytoplasm are processed similar to mRNA and transported out of the nucleus, while others with only one or very few exons are transported through the nuclear RNA export factor 1 (NXF1) [21].

## 4. Functions of lncRNAs

Although lncRNAs are being studied for the last two decades, still sufficient information needs to be gathered as compared to other non-coding RNAs. However, recent researches show that it has a role in multiple biological processes. It has been shown to function at multiple levels including regulation at the transcriptional and post-transcriptional level, structural function, and had roles at the level of genome integrity. Here, we are going to describe the role of a few well-studied lncRNAs in little detail and tabulate different lncRNAs with their known functions. The lncRNAs functions can be understood in the following ways:

### 4.1 Regulation at transcriptional level

To understand the regulatory role of lncRNAs at the transcription level, it is best to use the example of well characterized lncRNA, Xist, which is most studied among others. Xist is a ~ 17 kb long lncRNA, which is synthesized and expressed from X-chromosome (inactive state) and represses the gene expression through PRC1 and PRC2 [22–25]. As we all are aware that the female mammals carry a pair of X chromosome, while males carry single X chromosome, therefore one of the X chromosome is inactivated in females during early developmental events to ensure the dosage compensation between the two genders and Xist plays an important role in the process of X Chromosome Inactivation (XCI) [26]. Xist helps in maintaining the 3-D conformation of the X-chromosome such that it appears to be fully compact and maintains its inactive state (Xi) i.e. heterochromatin structure [26]. When Xist gets depleted from the chromatin, the inactive state of the X-chromosome gets its active state (Xa) by the process of X chromosome reactivation (XCR) which is not completely understood. How exactly does Xist play its role is through various chromatin factors including BRG1 and cohesin get repelled from the inactive state of the X-chromosome, leading to the disruption of the topologically associated domain (TADs) and in turn preventing the formation of chromatin super loops [27, 28]. In order to make the inactive state (the higher-order heterochromatin structure) Xist accompanies PRC1, PRC2 (Polycomb Repressive Complexes), and SMCHD1 (Structural Maintenance of Chromosomes Flexible Hinge Domain Containing 1) [29]. Altogether, it regulates gene expression by recruiting epigenetic factors or functioning as a protein complex scaffold. **Figure 3** represents the entire process diagrammatically.

HOTAIR, another lincRNA, which is found to work at transcriptional level. LincRNA, stands for long intergenic non-coding RNA, a sub class of long non-coding RNA (lncRNA). LincRNAs are transcribed from the intergenic regions of protein-coding genes. HOTAIR, one among many lincRNAs, is synthesized from the genomic region of HOXC gene and it is of ~2.2 kb in length. It controls the gene expression by modulating histone modification of target gene at HOXD loci [31]. The lincRNA binds to Polycomb Repressive Complex 2 (PRC2) and silence the transcription of HOXD

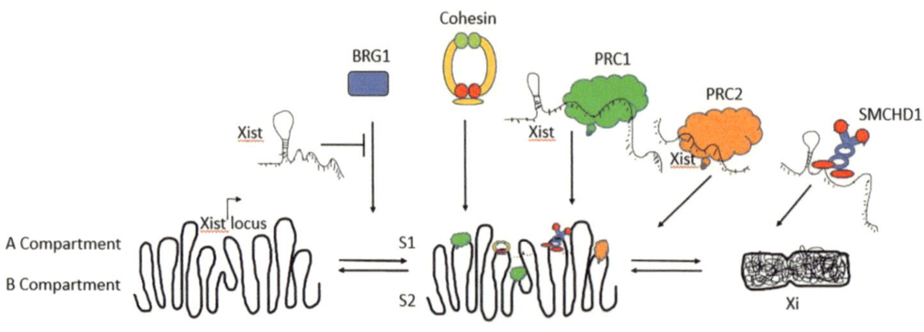


**Figure 3.**
*Regulatory roles of Xist. Xist inhibits the BRG1 and Cohesin, and recruits PRC1 and PRC2 to fold in compact form. Later it tethers with SMCHD1 to further compact the chromosome to form Xi. Adapted from [30].*

loci [31–33]. The large number of lncRNAs are now discovered because of technological advancement and found to function as HOTAIR [10, 34–36]. It is also found to be associated with cancer and play important role in tumorigenesis, and metastasis [34].

There are many lncRNAs are listed in **Table 1** that function similarly or differently at the transcriptional level and control the gene expression and cell fate. Some of the lncRNAs AIRN [80], ANCR [81, 82], ANRIL [37, 83, 84], BCAR4 [85] are nuclearly localized. AIRN functions at transcriptional level while remaining three controls the gene expression by modulating the histone modification.

### 4.2 Regulation at post-transcriptional level

There are lncRNAs which control the gene expression at post transcriptional levels. Here, PNUTS lncRNA serve the purpose to understand the mechanism they use. PNUTS is also known as PPP1R10, which generates mRNA, but alternative splicing leads to the synthesis of PNUTS lncRNA. This lncRNA actually functions as sponge and have binding sites for mir-205, which has been shown to bind ZEB1 and ZEB2 mRNA and causes the degradation of ZEB1 and ZEB2 mRNA. ZEB1 and ZEB2 are well known transcription factors associated with epithelial-mesenchymal transition [55]. In normal condition, the level of ZEB1 and ZEB2 is regulated by mir-205, but in cancer condition, when PNUTS lncRNA level goes up, mir-205 competitively binds to the lncRNA and becomes unavailable to their target ZEB1 and ZEB2, therefore it cannot degrade ZEB1 and ZEB2 mRNA, ultimately leading to the high level of ZEB1 and ZEB2 transcription factors. In turn, EMT proceeds and cancer progresses. **Figure 4** illustrates the role of PNUTS as post-transcriptional regulator of gene expression. This lncRNA does not affect the transcription process, but it does regulation of gene expression through microRNA-sponge function [55].

MALAT1 stands for Metastasis-associated lung adenocarcinoma transcript 1, also known as Nuclear Enriched Abundant Transcript 2 (NEAT2). It is ~8 kb long and transcribed from single exon [86]. As the name suggests, MALAT1 was first identified as metastasis associated lncRNA in non-small cell lung cancer (NSCLC) and was used as prognostic marker for NSCLC patient [87]. MALAT1 plays important role in splicing of mRNA. It interacts with serine-arginine splicing factor (SR protein) and governs the distribution of various splicing factors in nuclear speckle domain. It maintains the phosphorylated SR proteins level and changes in the level of MALAT1 affect the alternate splicing of endogenous mRNAs [88].

There are many other lncRNAs (See the **Table 1**), which work similar to PNUTS lncRNA and control the expression of different genes and control many biological processes.

### 4.3 Structural roles of lncRNAs

MALAT1 is so far well characterized lncRNA. The lncRNA has been shown to function at different levels including transcriptional where it facilitates the transcription factor binding to the promoters, can be part of splicing regulation, can regulate the gene expression epigenetically by interacting with PRC2 components namely EZH2, EED and SUZ12 to block miRNA or other gene expression usually through trimethylation at lysine 27 of histone H3 [89]. The lncRNA can be used as biomarkers and this can be chosen for targeted drug therapy. MALAT1 also functions as sponge for miR-1 binding and found to be engaged in the development of

| Function | lncRNAs | Interacting Partner | Mechanism of Function | Patho-physiology | References |
|---|---|---|---|---|---|
| Regulation of Transcription | ANRIL | PRC1, PRC2 | Recruits PRC to the promoters of CDKN2A and CDKN2B | Cancer and other diseases | [37, 38] |
| | LINC-PINT | PRC2 | Suppresses the gene expression | Down-regulated in many cancers | [39] |
| | lncPRESS1 | Sirtuin 6 | As decoy to regulate gene expression | ESC differentiation | [40] |
| | Xist | PRC2, hnRNPK, YY1 | Inactivate gene on X-chromosome | Cancer and development | [41–46] |
| | TARID | GADD45A | Forms R-loops | Demethylation | [47, 48] |
| | UMLILO | WDR5-MLL | CXCL chemokines transcription | Transcription of immune genes | [49] |
| | HOTTIP | WDR5-MLL | HOXA genes | Leukaemogenesis | [50, 51] |
| | COOLAIR | PRC2 | Histone H3 K27-me3 | Regulates flowering time | [52, 53] |
| Post-transcriptional Regulation | PNCTR | PTBP1 | Inhibits splicing | Upregulated in cancers | [54] |
| | PNUTS | mir-205 | Upregulate ZEB1 & ZEB2, promote EMT | EMT in Breast Cancer | [55] |
| | TINCR | STAU1 | RNA stability and expression | Dysregulated in many cancers | [56] |
| | FAST | β-TrCP | Inhibit β-catenin degradation, activate WNT signaling | Pluripotency | [13] |
| | NKILA | p65 | Inhibits NF-κB | Silencing of NKILA improves immune therapy | [57] |
| Structural Functions | NEAT1 | MALAT1 | Scaffold lncRNA | Breast and Skin cancers | [58–63] |
| | MALAT1 | NEAT1, U1 snRNA, SR protein | SR protein phosphorylation | Expressed in many cancers | [58, 63–71] |
| | sno-lncRNAs | RBFOX2 | mRNA splicing | Prader-Willi Syndrome | [72] |
| Genome Integrity | lincRNA-p21 | hnRNPK | Repress p53 induced gene expression | Dysregulated in many cancers | [73, 74] |
| | PANDA | NF-YA | Repress proapoptotic gene | Inhibits apoptosis and senescence | [75] |
| | NORAD | Pumilio, RBMX | Promote genomic stability | Dysregulated in many cancers | [76–79] |

**Table 1.**
*List of lncRNAs with their functions.*

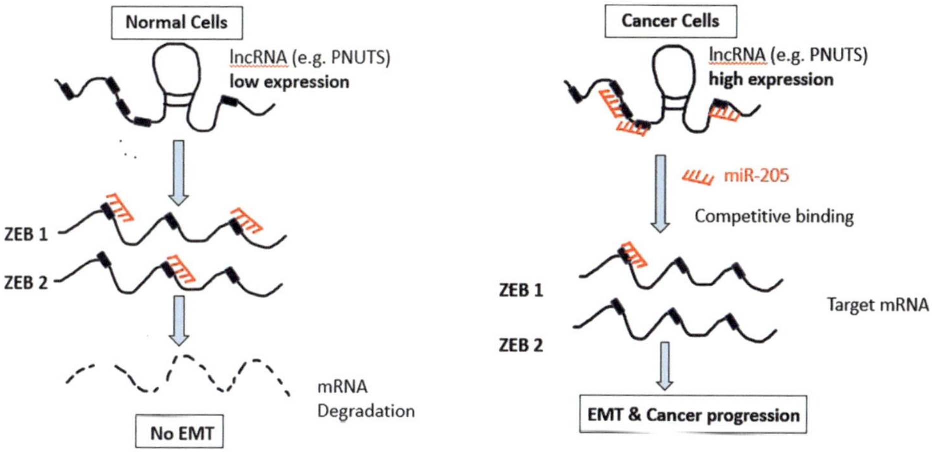


**Figure 4.**
*PNUTs as miRNA sponge. Regulation of Zeb1 and Zeb2 expression at post-transcriptional level. The level of PNUTS affect the EMT process. In cancer cells, PNUTS is expressed at high level and functions as miRNA sponge to bind miR-205 which targets ZEB1 and ZEB2, and EMT proceeds. In normal cells miR-205 binds to ZEB1 and ZEB2 and degrade them, ultimately results in No EMT progression.*

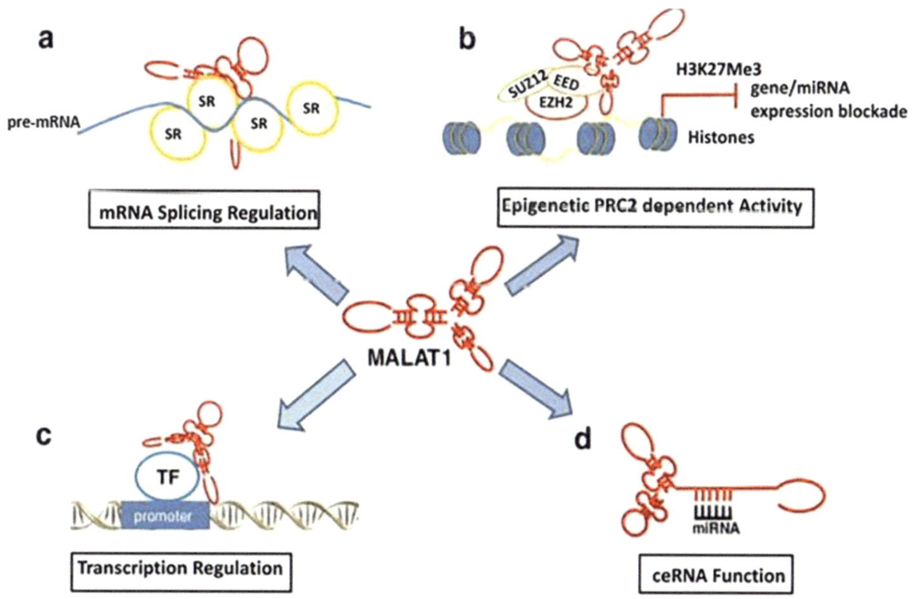


**Figure 5.**
*Regulatory roles of MALAT1. MALAT1 (~8.7kb IcnRNA, single exon) governs the distribution of splicing factors and play an important role in alternate splicing (a). It also control the gene expression via epigenetic machinery using PRC2 (b). It regulates the transcription (c) and can effectively trap miRNAs as sponge (d).*

bone and joint diseases [90]. MALAT1 functions as miRNA sponge and captures miR-1, which is associated with Cx43 repression and OPLL (Ossification of the posterior longitudinal ligament). **Figure 5** represents the explained roles of lncRNA MALAT1.

### 4.4 Other roles of lncRNAs

lncRNAs have been shown to perform various function. It interacts with all major biomolecules such as DNA, RNA, and proteins and modulate chromatin remodeling, expression of neighboring (adjacent or nearby) or distant genes. It can also affect RNA splicing, RNA stability and translation [91]. It directly interacts with DNA to form R-loop or RNA DNA triplex (RNA: DNA: DNA loop). It functions to control and regulate the gene expression at chromatin level to modulate the histone modification and activation or repression of gene. It also functions as sponge to capture multiple miRNAs and ultimately governs the expression of genes [92]. It also functions in rRNA maturation in mitochondria, a lncRNA RMRP is a part of mitochondrial RNA processing endoribonuclease (MRP) and carry out the maturation of rRNA [93].

## 5. Conclusions

lncRNAs are a comparatively new class of non-coding RNA, which has been shown to execute many biological functions. The lncRNA genes can be found anywhere in the genome e.g. intronic, overlapping, anti-sense, or stand-alone. It has been shown to perform many biological functions including a regulatory role in controlling gene expression. This sub-class of ncRNAs is tissue-specific and often shows differential expression patterns in diseases including cancers and heart diseases. These lncRNAs can be utilized as biomarkers as well as for targeted therapy. They function to regulate gene expression at various levels such as transcriptional and post-transcriptional as well as structural. It also functions as a sponge for miRNAs binding and adds another way of regulating gene expression. They may be a key player in cancer progression. It needs further investigations to find its involvement in other biological functions. This will help us to move forward from considering junk to useful biomolecules. The current understanding of lncRNA biology is somewhat limited, which will be further discussed and elaborated on in the future.


## Acknowledgements

I would like to express my sincere thanks to the Indian Institute of Technology Dharwad for providing all the necessary facilities and access to different journals. I am thankful to Paula Gavran for her help and assistance throughout the submission of this chapter.


## Authors' contribution

PK drafted the chapter, edited the figures as per reviewers' comments and suggestions. NB drew the figures. The authors read and approved the chapter for submission.

## Conflict of interest

The authors declare no conflict of interest.

## Author details

Pranjal Kumar* and Nikita Bhandari
Department of Biosciences and Bioengineering, Indian Institute of Technology Dharwad, Dharwad, Karnataka, India

*Address all correspondence to: 183021001@iitdh.ac.in; pranjal.dewangan6788@gmail.com

## References

[1] Ponting CP, Oliver PL, Reik W. Evolution and functions of long noncoding RNAs. Cell. 2009;**136**(4): 629-641

[2] Derrien T, Johnson R, Bussotti G, Tanzer A, Djebali S, Tilgner H, et al. The GENCODE v7 catalog of human long noncoding RNAs: Analysis of their gene structure, evolution, and expression. Genome Research. 2012;**22**(9):1775-1789

[3] Uszczynska-Ratajczak B, Lagarde J, Frankish A, Guigó R, Johnson R. Towards a complete map of the human long non-coding RNA transcriptome. Nature Reviews Genetics. 2018;**19**(9):535-548

[4] Fang S, Zhang L, Guo J, Niu Y, Wu Y, Li H, et al. NONCODEV5: A comprehensive annotation database for long non-coding RNAs. Nucleic Acids Research. 2017;**46**(D1):D308-D314

[5] Guttman M, Amit I, Garber M, French C, Lin MF, Feldser D, et al. Chromatin signature reveals over a thousand highly conserved large non-coding RNAs in mammals. Nature. 2009;**458**(7235):223-227

[6] Kung JTY, Colognori D, Lee JT. Long noncoding RNAs: Past, present, and future. Genetics. 2013;**193**(3):651-669

[7] Schmitt AM, Chang HY. Long noncoding RNAs: At the intersection of cancer and chromatin biology. Cold Spring Harbor Perspectives in Medicine. 2017;7(7):a026492

[8] Schmitz SU, Grote P, Herrmann BG. Mechanisms of long noncoding RNA function in development and disease. Cellular and Molecular Life Sciences. 2016;**73**(13):2491-2509

[9] Turton N, Swan R, Mahenthiralingam T, Pitts D, Dykes IM. The functions of long non-coding RNA during embryonic cardiovascular development and its potential for diagnosis and treatment of congenital heart disease. Journal of Cardiovascular Development and Disease. 2019;**6**(2):21

[10] Bhan A, Mandal SS. LncRNA HOTAIR: A master regulator of chromatin dynamics and cancer. Biochimica et Biophysica Acta (BBA) - Reviews on Cancer. 2015;**1856**(1):151-164

[11] International Human Genome Sequencing Consortium. Initial sequencing and analysis of the human genome. Nature. 2001;**409**(6822):860-921

[12] Venter JC, Adams MD, Myers EW, Li PW, Mural RJ, Sutton GG, et al. The sequence of the human genome. Science. 2001;**291**(5507):1304-1351

[13] Guo C-J, Ma X-K, Xing Y-H, Zheng C-C, Xu Y-F, Shan L, et al. Distinct processing of lncRNAs contributes to non-conserved functions in stem cells. Cell. 2020;**181**(3):621-636.e22

[14] Quinn JJ, Zhang QC, Georgiev P, Ilik IA, Akhtar A, Chang HY. Rapid evolutionary turnover underlies conserved lncRNA–genome interactions. Genes & Development. 2016;**30**(2):191-207

[15] Hezroni H, Koppstein D, Schwartz Matthew G, Avrutin A, Bartel David P, Ulitsky I. Principles of long noncoding RNA evolution derived from direct comparison of transcriptomes in 17 species. Cell Reports. 2015;**11**(7):1110-1122

[16] Lagarde J, Uszczynska-Ratajczak B, Carbonell S, Pérez-Lluch S, Abad A, Davis C, et al. High-throughput annotation of full-length long noncoding RNAs with capture long-read sequencing. Nature Genetics. 2017;**49**(12):1731-1740

[17] Melé M, Mattioli K, Mallard W, Shechner DM, Gerhardinger C, Rinn JL. Chromatin environment, transcriptional regulation, and splicing distinguish lincRNAs and mRNAs. Genome Research. 2016;**27**(1):27-37

[18] Schlackow M, Nojima T, Gomes T, Dhir A, Carmo-Fonseca M, Proudfoot NJ. Distinctive patterns of transcription and RNA processing for human lincRNAs. Molecular Cell. 2017;**65**(1):25-38

[19] Rosenberg Alexander B, Patwardhan Rupali P, Shendure J, Seelig G. Learning the sequence determinants of alternative splicing from millions of random sequences. Cell. 2015;**163**(3):698-711

[20] Zuckerman B, Ulitsky I. Predictive models of subcellular localization of long RNAs. RNA. 2019;**25**(5):557-572

[21] Zuckerman B, Ron M, Mikl M, Segal E, Ulitsky I. Gene architecture and sequence composition underpin selective dependency of nuclear export of long RNAs on NXF1 and the TREX complex. Molecular Cell. 2020;**79**(2):251-267.e6

[22] Brown CJ, Hendrich BD, Rupert JL, Lafrenière RG, Xing Y, Lawrence J, et al. The human XIST gene: Analysis of a 17 kb inactive X-specific RNA that contains conserved repeats and is highly localized within the nucleus. Cell. 1992;**71**(3):527-542

[23] Wang J, Mager J, Chen Y, Schneider E, Cross JC, Nagy A, et al. Imprinted X inactivation maintained by a mouse Polycomb group gene. Nature Genetics. 2001;**28**(4):371-375

[24] Plath K, Talbot D, Hamer KM, Otte AP, Yang TP, Jaenisch R, et al. Developmentally regulated alterations in Polycomb repressive complex 1 proteins on the inactive X chromosome. Journal of Cell Biology. 2004;**167**(6):1025-1035

[25] Zhao J, Sun BK, Erwin JA, Song J-J, Lee JT. Polycomb proteins targeted by a short repeat RNA to the mouse X chromosome. Science. 2008;**322**(5902):750-756

[26] Dixon McDougall T, Brown C. The making of a Barr body: The mosaic of factors that eXIST on the mammalian inactive X chromosome. Biochemistry and Cell Biology. 2016;**94**(1):56-70

[27] Minajigi A, Froberg JE, Wei C, Sunwoo H, Kesner B, Colognori D, et al. A comprehensive Xist interactome reveals cohesin repulsion and an RNA-directed chromosome conformation. Science. 2015;**349**(6245)

[28] Jégu T, Blum R, Cochrane JC, Yang L, Wang C-Y, Gilles M-E, et al. Xist RNA antagonizes the SWI/SNF chromatin remodeler BRG1 on the inactive X chromosome. Nature Structural & Molecular Biology. 2019;**26**(2):96-109

[29] Wang C-Y, Jégu T, Chu H-P, Oh HJ, Lee JT. SMCHD1 merges chromosome compartments and assists formation of super-structures on the inactive X. Cell. 2018;**174**(2):406-421.e25

[30] Guh C-Y, Hsieh Y-H, Chu H-P. Functions and properties of nuclear lncRNAs—From systematically mapping the interactomes of lncRNAs. Journal of Biomedical Science. 2020;**27**(1):44

[31] Rinn JL, Kertesz M, Wang JK, Squazzo SL, Xu X, Brugmann SA, et al. Functional demarcation of active and silent chromatin domains in human

HOX loci by noncoding RNAs. Cell. 2007;**129**(7):1311-1323

[32] Kopp F, Mendell JT. Functional classification and experimental dissection of long noncoding RNAs. Cell. 2018;**172**(3):393-407

[33] Portoso M, Ragazzini R, Brenčič Ž, Moiani A, Michaud A, Vassilev I, et al. PRC2 is dispensable for *HOTAIR*-mediated transcriptional repression. The EMBO Journal. 2017;**36**(8):981-994

[34] Tang Q, Hann S. HOTAIR: An oncogenic long non-coding RNA in human cancer. Cellular Physiology and Biochemistry. 2018;**47**(3):893-913

[35] Zhou X, Chen J, Tang W. The molecular mechanism of HOTAIR in tumorigenesis, metastasis, and drug resistance. Acta Biochimica et Biophysica Sinica. 2014;**46**(12):1011-1015

[36] Wu Y, Zhang L, Wang Y, Li H, Ren X, Wei F, et al. Long noncoding RNA HOTAIR involvement in cancer. Tumor Biology. 2014;**35**(10):9531-9538

[37] Yap KL, Li S, Muñoz-Cabello AM, Raguz S, Zeng L, Mujtaba S, et al. Molecular interplay of the noncoding RNA ANRIL and methylated histone H3 lysine 27 by Polycomb CBX7 in transcriptional silencing of INK4a. Molecular Cell. 2010;**38**(5):662-674

[38] Holdt LM, Hoffmann S, Sass K, Langenberger D, Scholz M, Krohn K, et al. Alu elements in ANRIL non-coding RNA at chromosome 9p21 modulate Atherogenic cell functions through trans-regulation of gene networks. McCarthy MI, editor. PLoS Genetics. 2013;**9**(7):e1003588

[39] Marín-Béjar O, Mas AM, González J, Martinez D, Athie A, Morales X, et al. The human lncRNA LINC-PINT inhibits tumor cell invasion through a highly conserved sequence element. Genome Biology. 2017;**18**(1):202

[40] Jain AK, Xi Y, McCarthy R, Allton K, Akdemir KC, Patel LR, et al. LncPRESS1 is a p53-regulated LncRNA that safeguards pluripotency by disrupting SIRT6-mediated De-acetylation of histone H3K56. Molecular Cell. 2016;**64**(5):967-981

[41] Schertzer MD, Braceros KCA, Starmer J, Cherney RE, Lee DM, Salazar G, et al. lncRNA-induced spread of Polycomb controlled by genome architecture, RNA abundance, and CpG Island DNA. Molecular Cell. 2019;**75**(3):523-537.e10

[42] Lee Jeannie T, Bartolomei MS. X-inactivation, imprinting, and long noncoding RNAs in health and disease. Cell. 2013;**152**(6):1308-1323

[43] Pintacuda G, Wei G, Roustan C, Kirmizitas BA, Solcan N, Cerase A, et al. hnRNPK recruits PCGF3/5-PRC1 to the Xist RNA B-repeat to establish Polycomb-mediated chromosomal silencing. Molecular Cell. 2017;**68**(5):955-969.e10

[44] Colognori D, Sunwoo H, Kriz AJ, Wang C-Y, Lee JT. Xist Deletional analysis reveals an interdependency between Xist RNA and Polycomb complexes for spreading along the inactive X. Molecular Cell. 2019;**74**(1):101-117.e10

[45] McHugh CA, Chen C-K, Chow A, Surka CF, Tran C, McDonel P, et al. The Xist lncRNA interacts directly with SHARP to silence transcription through HDAC3. Nature. 2015;**521**(7551):232-236

[46] Jeon Y, Lee JT. YY1 tethers Xist RNA to the inactive X nucleation center. Cell. 2011;**146**(1):119-133

[47] Arab K, Karaulanov E, Musheev M, Trnka P, Schäfer A, Grummt I, et al. GADD45A binds R-loops and recruits TET1 to CpG island promoters. Nature Genetics. 2019;**51**(2):217-223

[48] Arab K, Park YJ, Lindroth AM, Schäfer A, Oakes C, Weichenhan D, et al. Long noncoding RNA TARID directs demethylation and activation of the tumor suppressor TCF21 via GADD45A. Molecular Cell. 2014;**55**(4):604-614

[49] Fanucchi S, Fok ET, Dalla E, Shibayama Y, Börner K, Chang EY, et al. Immune genes are primed for robust transcription by proximal long noncoding RNAs located in nuclear compartments. Nature Genetics. 2018;**51**(1):138-150

[50] Wang KC, Yang YW, Liu B, Sanyal A, Corces-Zimmerman R, Chen Y, et al. A long noncoding RNA maintains active chromatin to coordinate homeotic gene expression. Nature. 2011;**472**(7341):120-124

[51] Luo H, Zhu G, Xu J, Lai Q, Yan B, Guo Y, et al. HOTTIP lncRNA promotes hematopoietic stem cell self-renewal leading to AML-like disease in mice. Cancer Cell. 2019;**36**(6):645-659.e8

[52] Rosa S, Duncan S, Dean C. Mutually exclusive sense–antisense transcription at FLC facilitates environmentally induced gene repression. Nature Communications. 2016;**7**(1):13031

[53] Csorba T, Questa JI, Sun Q, Dean C. Antisense COOLAIR mediates the coordinated switching of chromatin states at FLC during vernalization. Proceedings of the National Academy of Sciences. 2014;**111**(45):16160-16165

[54] Yap K, Mukhina S, Zhang G, Tan JSC, Ong HS, Makeyev EV. A short tandem repeat-enriched RNA assembles a nuclear compartment to control alternative splicing and promote cell survival. Molecular Cell. 2018, 72;(3):525-540.e13

[55] Grelet S, Link LA, Howley B, Obellianne C, Palanisamy V, Gangaraju VK, et al. A regulated PNUTS mRNA to lncRNA splice switch mediates EMT and tumour progression. Nature Cell Biology. 2017;**19**(9): 1105-1115

[56] Kretz M, Siprashvili Z, Chu C, Webster DE, Zehnder A, Qu K, et al. Control of somatic tissue differentiation by the long non-coding RNA TINCR. Nature. 2012;**493**(7431):231-235

[57] Huang D, Chen J, Yang L, Ouyang Q, Li J, Lao L, et al. NKILA lncRNA promotes tumor immune evasion by sensitizing T cells to activation-induced cell death. Nature Immunology. 2018;**19**(10):1112-1125

[58] Hutchinson JN, Ensminger AW, Clemson CM, Lynch CR, Lawrence JB, Chess A. A screen for nuclear transcripts identifies two linked noncoding RNAs associated with SC35 splicing domains. BMC Genomics. 2007;**8**:39

[59] Sasaki YTF, Ideue T, Sano M, Mituyama T, Hirose T. MENε/β noncoding RNAs are essential for structural integrity of nuclear paraspeckles. Proceedings of the National Academy of Sciences. 2009;**106**(8):2525-2530

[60] Sunwoo H, Dinger ME, Wilusz JE, Amaral PP, Mattick JS, Spector DL. MEN ε/β nuclear-retained non-coding RNAs are up-regulated upon muscle differentiation and are essential components of paraspeckles. Genome Research. 2008;**19**(3):347-359

[61] Yamazaki T, Souquere S, Chujo T, Kobelke S, Chong YS, Fox AH,

et al. Functional domains of NEAT1 architectural lncRNA induce Paraspeckle assembly through phase separation. Molecular Cell. 2018;**70**(6):1038-1053.e7

[62] Lin Y, Schmidt BF, Bruchez MP, McManus C. Structural analyses of NEAT1 lncRNAs suggest long-range RNA interactions that may contribute to paraspeckle architecture. Nucleic Acids Research. 2018;**46**(7):3742-3752

[63] Lu Z, Zhang Q, Lee B, Flynn Ryan A, Smith Martin A, Robinson James T, et al. RNA duplex map in living cells reveals higher-order transcriptome structure. Cell. 2016;**165**(5):1267-1279

[64] Wilusz JE, JnBaptiste CK, Lu LY, Kuhn C-D, Joshua-Tor L, Sharp PA. A triple helix stabilizes the 3′ ends of long noncoding RNAs that lack poly(a) tails. Genes & Development. 2012;**26**(21):2392-2407

[65] Tripathi V, Song DY, Zong X, Shevtsov SP, Hearn S, Fu X-D, et al. SRSF1 regulates the assembly of pre-mRNA processing factors in nuclear speckles. Weis K, editor. Molecular Biology of the Cell. 2012;**23**(18):3694-3706

[66] Yang L, Lin C, Liu W, Zhang J, Ohgi Kenneth A, Grinstein Jonathan D, et al. ncRNA- and Pc2 methylation-dependent gene relocation between nuclear structures mediates gene activation programs. Cell. 2011;**147**(4):773-788

[67] Arun G, Diermeier S, Akerman M, Chang K-C, Wilkinson JE, Hearn S, et al. Differentiation of mammary tumors and reduction in metastasis upon *Malat1* lncRNA loss. Genes & Development. 2015;**30**(1):34-51

[68] Malakar P, Shilo A, Mogilevsky A, Stein I, Pikarsky E, Nevo Y, et al. Long noncoding RNA MALAT1 promotes hepatocellular carcinoma development by SRSF1 upregulation and mTOR activation. Cancer Research. 2016;**77**(5):1155-1167

[69] Fei J, Jadaliha M, Harmon TS, Li ITS, Hua B, Hao Q, et al. Quantitative analysis of multilayer organization of proteins and RNA in nuclear speckles at super resolution. Journal of Cell Science. 2017;**130**(24):4180-4192

[70] Engreitz Jesse M, Sirokman K, McDonel P, Shishkin AA, Surka C, Russell P, et al. RNA-RNA interactions enable specific targeting of noncoding RNAs to nascent pre-mRNAs and chromatin sites. Cell. 2014;**159**(1):188-199

[71] Wilusz JE, Freier SM, Spector DL. 3′ end processing of a long nuclear-retained noncoding RNA yields a tRNA-like cytoplasmic RNA. Cell. 2008;**135**(5):919-932

[72] Yin Q-F, Yang L, Zhang Y, Xiang J-F, Wu Y-W, Carmichael Gordon G, et al. Long noncoding RNAs with snoRNA ends. Molecular Cell. 2012;**48**(2):219-230

[73] Huarte M, Guttman M, Feldser D, Garber M, Koziol MJ, Kenzelmann-Broz D, et al. A large intergenic noncoding RNA induced by p53 mediates global gene repression in the p53 response. Cell. 2010;**142**(3):409-419

[74] Dimitrova N, Zamudio Jesse R, Jong Robyn M, Soukup D, Resnick R, Sarma K, et al. LincRNA-p21 activates p21 In cis to promote Polycomb target gene expression and to enforce the G1/S checkpoint. Molecular Cell. 2014;**54**(5):777-790

[75] Hung T, Wang Y, Lin MF, Koegel AK, Kotake Y, Grant GD, et al.

Extensive and coordinated transcription of noncoding RNAs within cell cycle promoters. Nature Genetics. 2011;**43**(7):621-629

[76] Lee S, Kopp F, Chang T-C, Sataluri A, Chen B, Sivakumar S, et al. Noncoding RNA NORAD regulates genomic stability by sequestering PUMILIO proteins. Cell. 2016;**164**(1-2):69-80

[77] Tichon A, Perry RB-T, Stojic L, Ulitsky I. SAM68 is required for regulation of Pumilio by the NORAD long noncoding RNA. Genes & Development. 2018;**32**(1):70-78

[78] Schmitt AM, Garcia JT, Hung T, Flynn RA, Shen Y, Qu K, et al. An inducible long noncoding RNA amplifies DNA damage signaling. Nature Genetics. 2016;**48**(11):1370-1376

[79] Munschauer M, Nguyen CT, Sirokman K, Hartigan CR, Hogstrom L, Engreitz JM, et al. The NORAD lncRNA assembles a topoisomerase complex critical for genome stability. Nature. 2018;**561**(7721):132-136

[80] Latos PA, Pauler FM, Koerner MV, Şenergin HB, Hudson QJ, Stocsits RR, et al. *Airn* transcriptional overlap, but not its lncRNA products, induces imprinted *Igf2r* silencing. Science. 2012;**338**(6113):1469-1472

[81] Kretz M, Webster DE, Flockhart RJ, Lee CS, Zehnder A, Lopez-Pajares V, et al. Suppression of progenitor differentiation requires the long noncoding RNA ANCR. Genes & Development. 2012;**26**(4):338-343

[82] Zhu L, Xu P-C. Downregulated LncRNA-ANCR promotes osteoblast differentiation by targeting EZH2 and regulating Runx2 expression. Biochemical and Biophysical Research Communications. 2013;**432**(4):612-617

[83] Pasmant E, Laurendeau I, Héron D, Vidaud M, Vidaud D, Bièche I. Characterization of a germ-line deletion, including the entire *INK4/ARF* locus, in a melanoma-neural system tumor family: Identification of *ANRIL*, an antisense noncoding RNA whose expression Coclusters with *ARF*. Cancer Research. 2007;**67**(8):3963-3969

[84] Visel A, Zhu Y, May D, Afzal V, Gong E, Attanasio C, et al. Targeted deletion of the 9p21 non-coding coronary artery disease risk interval in mice. Nature. 2010;**464**(7287): 409-412

[85] Xing Z, Lin A, Li C, Liang K, Wang S, Liu Y, et al. lncRNA directs cooperative epigenetic regulation downstream of chemokine signals. Cell. 2014;**159**(5):1110-1125

[86] Zhao Y, Yu Y, You S, Zhang C, Wu L, Zhao W, et al. Long non-coding RNA MALAT1 as a detection and diagnostic molecular marker in various human cancers: A pooled analysis based on 3255 subjects. OncoTargets and Therapy. 2020;**13**:5807-5817

[87] Sun Y, Ma L. New insights into long non-coding RNA MALAT1 in cancer and metastasis. Cancers. 2019;**11**(2):216

[88] Tripathi V, Ellis JD, Shen Z, Song DY, Pan Q, Watt AT, et al. The nuclear-retained noncoding RNA MALAT1 regulates alternative splicing by modulating SR splicing factor phosphorylation. Molecular Cell. 2010;**39**(6):925-938

[89] Amodio N, Raimondi L, Juli G, Stamato MA, Caracciolo D, Tagliaferri P, et al. MALAT1: A druggable long non-coding RNA for targeted anti-cancer approaches. Journal of Hematology & Oncology. 2018;**11**(1):63

[90] Yuan X, Guo Y, Chen D, Luo Y, Chen D, Miao J, et al. Long non-coding RNA MALAT1 functions as miR-1 sponge to regulate Connexin 43-mediated ossification of the posterior longitudinal ligament. Bone. 2019;**127**:305-314

[91] Statello L, Guo C-J, Chen L-L, Huarte M. Gene regulation by long non-coding RNAs and its biological functions. Nature Reviews Molecular Cell Biology. 2021;**22**(2):96-118

[92] López-Urrutia E, Bustamante Montes LP. Ladrón de Guevara Cervantes D, Pérez-Plasencia C, Campos-Parra AD. Crosstalk between long non-coding RNAs, micro-RNAs and mRNAs: Deciphering molecular mechanisms of master regulators in cancer. Frontiers in. Oncology. 2019;**9**:669

[93] Noh JH, Kim KM, Abdelmohsen K, Yoon J-H, Panda AC, Munk R, et al. HuR and GRSF1 modulate the nuclear export and mitochondrial localization of the lncRNA *RMRP*. Genes & Development. 2016;**30**(10):1224-1239

Chapter 5

# Gene Expression and Transcriptome Sequencing: Basics, Analysis, Advances

*Nakul D. Magar, Priya Shah, K. Harish, Tejas C. Bosamia, Kalyani M. Barbadikar, Yogesh M. Shukla, Amol Phule, Harshvardhan N. Zala, Maganti Sheshu Madhav, Satendra Kumar Mangrauthia, Chirravuri Naga Neeraja and Raman Meenakshi Sundaram*

**Abstract**

Gene expression studies are extremely useful for understanding a broad range of biological, physiological, and molecular responses. The techniques for gene expression reflect differential patterns of gene regulation and have evolved with time from detecting one gene to many genes at a time laterally. Gene expression depends on the spatiotemporal expression in a particular tissue at a given time point and needs critical examination and interpretation. Transcriptome sequencing or RNA-seq using next-generation sequencing (short and long reads) is the most widely deployed technology for accurate quantification of gene expression. According to the biological aim of the experiment, replications, platform, and chemistries, propelling improvement has been demonstrated and documented using RNA-seq in plants, humans, animals, and clinical sciences with respect to gene expression of mRNA, small non-coding, long non-coding RNAs, alternative splice variations, isoform variations, gene fusions, single-nucleotide variants. Integrating transcriptome sequencing with other techniques such as chromatin immunoprecipitation, methylation, genome-wide association studies, manifests insights into genetic and epigenetic regulation. Epi-transcriptome including RNA methylation, modification, and alternative polyadenylation events can also be explored through long-read sequencing. In this chapter, we have presented an account of the basics of gene expression methods, transcriptome sequencing, and the various methodologies involved in the downstream analysis.

**Keywords:** ESTs, microarray, RNA-seq, assembly, annotation, visualization, tools, databases

## 1. Introduction

The phenotypic manifestation of the genetic code through transcription and translation is known as gene expression. The determination of specific spatiotemporal expression patterns under a particular condition or developmental stage is known as gene expression analysis. The gene expression analysis has gained feasible attention in the biological field of research. The conventional methods of gene expression and functional analysis focus on one gene at a time. But in the last decade, there has been the development of numerous high-throughput technologies that allow the expression studies of thousands of genes simultaneously, in a single experiment such as microarray, transcriptome analysis/ RNA-seq, etc. These methods are highly capable of generating an ample amount of biological data. There has been phenomenal progress in the data repositories, and the data are continuously being deposited in the databases. Parallelly, the advancements made in the bioinformatics pipeline, tools, and software (online/ offline) with the graphical user interface or language-based also add to the ease and convenience to use the same for data analysis. Several databases serve as repositories of the sequenced data, the most widely deployed is National Center for Biotechnology Information, NCBI.

## 2. Evolution of high-throughput transcriptomic technologies

The traditional methodologies for gene predictions and transcriptomic studies involve the complementary deoxyribonucleic acid (cDNA) clone preparation and further utilize it to generate expressed sequence tags (ESTs), and then sequencing these tags using the first-generation sequencing platforms such as Sanger sequencing technology. **Figure 1** shows historic advancements in gene expression technologies. In the late 1990s, gene expression studies were carried out for 45 *Arabidopsis* genes by using the early high-capacity microarrays in which cDNA is spotted on microscope-sized glass slides [1]. Another pioneering quantitative transcriptomic study is the serial analysis of gene expression (SAGE), which was first performed on 1000 tags for characterization of the human pancreatic gene expression pattern [2]. Later on, with the advancement in sequencing technology, a technique such as RNA sequencing (RNA-seq) has emerged that possesses numerous next-generation

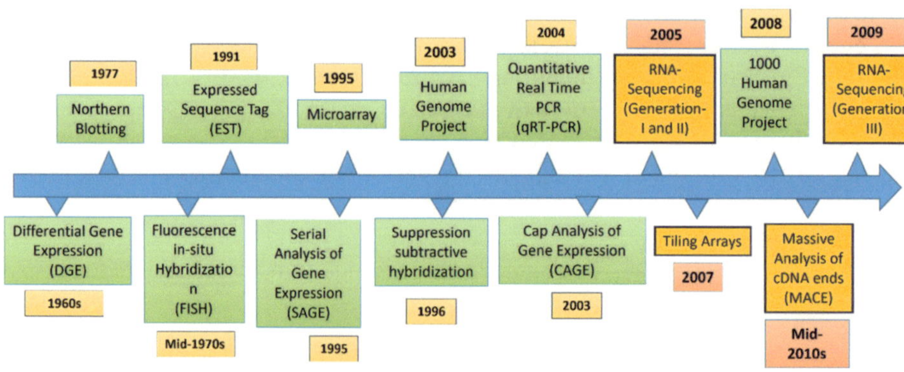


**Figure 1.**
*Historic timeline of technologies involved in gene expression analysis.*

sequencing techniques that help to retrieve the sequence and the expression level of the RNA transcripts [3, 4]. Continuous efforts have been made over the years for the development of feasible high-throughput technologies for gene expression profiling and quantification. This will help to cope with the several challenges associated with sequencing technology that include cost, complexity, availability, and error occurrence rate while assembling the sequence [5]. In comparison to the sequencing technology, the array-based technology does not involve these challenges, hence is still widely used for expression studies. However, it has several other limitations such as the probe-based nature of microarrays, it requires predefined probes, and hence, is unable to deliver precise readings [6].

With the onset of RNA sequencing (RNA-seq) technology, whole transcriptome sequencing has been carried out [7, 8]. RNA-seq studies cover the genome-wide assessment of transcripts and have a sequencing depth of 100–1000 reads per base pair of a transcript [9]. In the RNA-seq technology generally, the output comprises short reads, which are generated by sequencing the cDNA fragments from one end or both ends. Further, the error rate is minimized followed by assembling these short reads into the long sequences in correspondence to the sample RNAs.

Generally, for sequencing the short reads, the next-generation sequencing platforms are being utilized to read quite short sequences of 35–500 bp [5, 10]. This platform requires high-powered computing systems with huge storage and memory along with several cores as this will enable to run the algorithms simultaneously and regenerate the full-length transcripts. However, it has been observed that such platforms possess showcased coverage and have quite high error rates ultimately increasing the informatics challenges [6, 11, 12]. There would be a requirement for additional reads for ensuring high-quality coverage and improving throughput [9, 12]. The assembly algorithms have always kept evolving with time and improving the quality of data. The main aim is to read the extension of the length and eliminate the assembly dependency. The advanced RNA sequence technologies include single molecule, real-time sequencing technology (SMRT), or nanopore sequencers that can cope with the existing limitations and provide several kilobases longer reads and generate whole-genome transcripts. The SMRT platforms have an average read length of 3000 bp and are extendable up to 20,000 bp [9, 13, 14].

In combination with the fluorescent *in-situ* hybridization, RNA-seq technology has made an advancement in the data generation even at the transcript cellular localization. A cell's RNA is sequenced while it remains in tissue or culture using next-generation sequencing called fluorescent in situ sequencing (FISSEQ) [15] and is a breakthrough in transcriptomic research. In this technology, firstly the cDNA is generated by RNA reverse transcription in situ, then via rolling-circle amplification copies of cDNA are generated to form DNA "nanoballs." Then by making use of the "sequencing by oligonucleotide ligation and detection" (SOLiD) technology based on sequential hybridization of fluorescently labeled probes with two bases, these nanoballs are sequenced at the cellular level. The emergence of this technology has made possible the simultaneous generation of sequence and positional information. However, it still requires further optimization for wider adoption.

The novel sequencing approach such as nanopore sequencing can perform direct RNA sequencing by eliminating the need to generate cDNA and sequence assembly unlike several high-throughput technologies [16], by avoiding the dependency on the two important inherent sources of error in the application of indirect approaches. The most important point to take into account regarding all these methods from microarrays or next-generation sequencing is that when the simultaneous measurements have been carried out, despite a very low error rate, a large number of errors occur. Hence, there is a need for cross-validation to enhance the high-throughput data accuracy by utilizing an alternative procedure, such as quantitative real-time polymerase chain reaction (qRT-PCR) or other gene expression methods as discussed in the later part [17, 18].

## 3. Expressed sequenced tags (ESTs)

An EST is a short fragment of RNA sequence (200–800) generated from sequencing of randomly selected cDNA clones. Single RNA transcript is reverse transcribed to cDNA, cloned, then it is sequenced. These cDNA libraries will provide information on EST, which can be used to identify gene transcript, gene discovery, and sequence determination [19]. It involves mapping of EST to the location on a specific chromosome using physical mapping strategies or aligning EST sequence with the genome. It will help to find out the expression of the corresponding gene concerning specific conditions or any treatment [20]. Hence, ESTs are studying the structure of plant genome, gene expression, and function [21]. Additionally, this tool also helps to clarify the structural gene annotation and development of molecular markers [22, 23], genomic map construction [24], study ancestral relationships between the species, helps in the elucidation of transcriptome activity [25, 26] as well providing information to develop probes DNA chips [1]. **Figure 2** shows the methodology of EST-seq. With the advancement

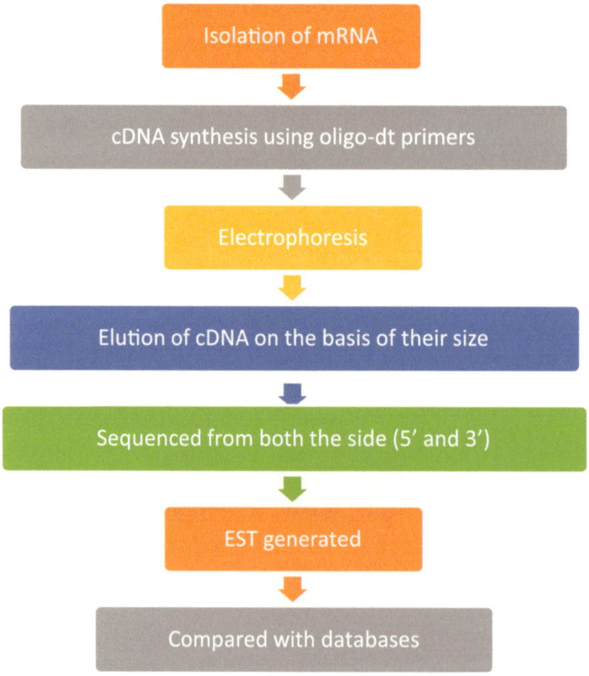


**Figure 2.**
*Flow chart of the EST sequencing.*

| S. No | Name of database | URL |
|---|---|---|
| 1 | dbEST at NCBI | http://www.ncbi.nlm.nih.gov/dbEST/index.html |
| 2 | DDBJ | http://ddbj.nig.ac.jp/index-e.html |
| 3 | EMBL-EBI | http://ebi.ac.uk/embl.html |
| 4 | Kazusa EST database | http://www.kazusa.or.jp/en/plant/database.html |
| 5 | Plant GDB | http://www.plantgdb.org/ |
| 6 | TIGR plant gene indices | http://www.tigr.org/tdb/tgi/plant.html |
| 7 | UniGene at NCBI | http://www.ncbi.nlm.nih.gov/UniGene/ |
| 8 | University Minnesota | http://ccgb.umm.edu/ |
| 9 | KGENES | http://www.genome.jp/kegg-bin/creare_kegg_ |
| 10 | ESTreedb | http://www.itb.cnr.it/estree |
| 11 | TbestDB | http://www.tbestdb.bcm.umontreal.ca |
| 12 | Pscroph database | http://www.pscroph.ucdavis.edu |
| 13 | Mendel-GFDb and Mendel-ESTS | http://www.mendel.ac.uk/ |
| 14 | US Mirror | http://genome.cornell.edu/ |
| 15 | Sputnik | http://www.mips.gsf.de/proj/sputni |

**Table 1.**
*Databases of ESTs.*

in sequencing techniques, various approaches such as whole-genome sequencing and transcriptome sequencing become an alternative for EST. These NGS techniques avoid the missing of rare transcripts by reducing the complexity and cost of sequencing [27]. Sanger sequencing method generated EST data with less number as compared with GS-FLX, which is being a widely used technique for *de novo* sequencing and EST analysis in plants [6]. **Table 1** shows the databases available for ESTs. The suppression subtractive hybridization (SSH) was developed for the generation of subtracted cDNA libraries based on suppression PCR. It combines normalization and subtraction in a single procedure wherein the common sequences between the two samples for differential gene expression are subtracted and the rare sequences are enriched [28]. The use of this technique is limited due to its complexity and to the identification of low abundance genes.

## 4. Serial analysis of gene expression (SAGE/CAGE)

Serial analysis of gene expression (SAGE) refers to the comprehensive, unbiased, and quantitative gene expression of transcript profiles. SAGE involves the development of EST with the help of high-throughput tags for quantification [2]. For several modifications for quantification, it does not require the prior knowledge gene, which is superseded over the array techniques. In SAGE, cDNA generated from respective mRNA is digested using specific restriction enzyme results into 10–11 bp tag fragments. Further, these tags are concatenated (head to tail) to long strands (>400–600 bp) and sequenced. The sequence is then aligned with the reference gene for the identification of corresponding gene (**Figure 3**). Lacking information on the reference genome, differentially expressed tags can also be used

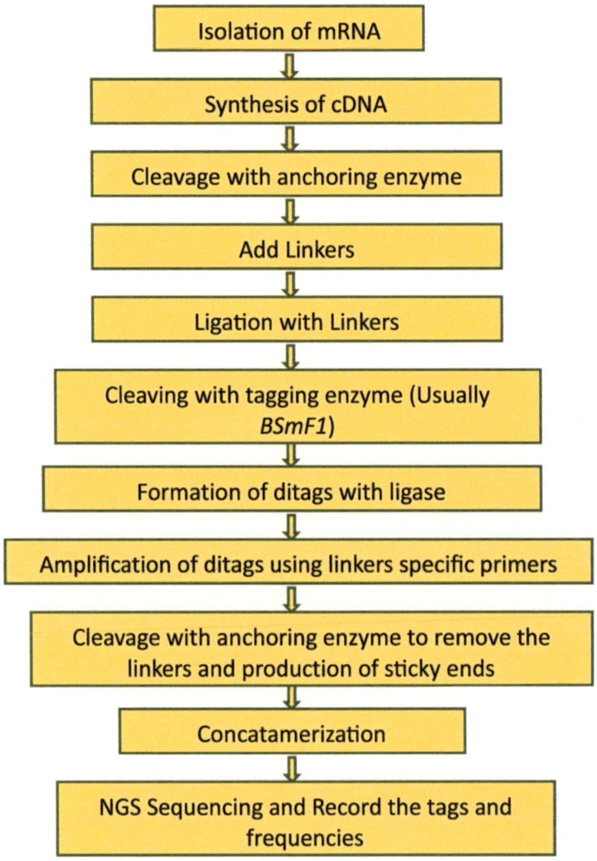


**Figure 3.**
*Flow chart for serial analysis of gene expression.*

as diagnostic markers. Variants of SAGE have been studied such as cap analysis of gene expression (CAGE) involves sequence tags from the 5A end of an mRNA transcript only [29]. Consequently, these tags aligned with the reference genome will help to reveal the transcriptional start site. Likewise, several SAGE-like variants have been developed (MAGE, SAGE, microSAGE, miniSAGE, longSAGE, superSAGE, deepSAGE, etc.) to study the genome-wide analysis of DNA copy-number changes and methylation patterns, chromatin structure, and transcription factor targets.

## 5. Microarray

For the last decade, for high-throughput transcriptome profiling, DNA microarrays have been preferred. The gene expression quantification requires RNA and microarrays hybridization. One such technique is the microarrays that rely on the principle of complementarity between the nucleic acid strands [30]. The microarrays are distinguished into two types: genotyping microarrays and expression microarrays. In the former, specific cDNA while the latter is used to detect specific RNA [31]. The comparison between the results of these two arrays leads

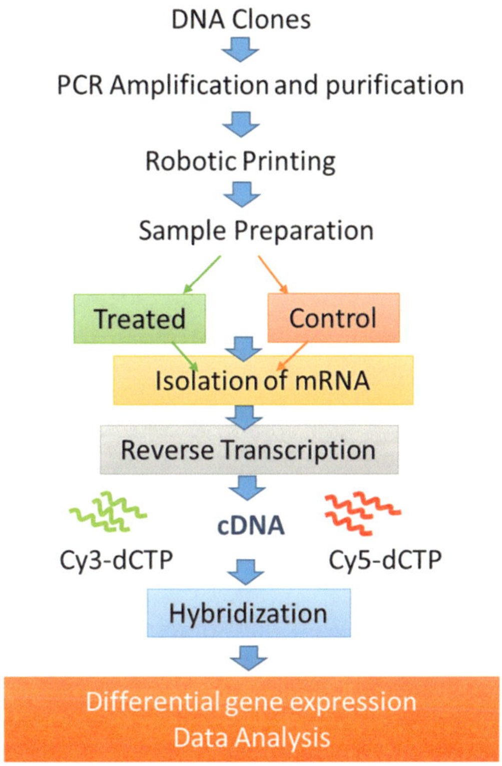


**Figure 4.**
*Flowchart for a methodology for microarray.*

to the establishment of the specific variation in the gene expression patterns and the mRNA abundance. This ultimately leads to the detection of some promising candidate genes in response to the different treatments and distinct genetic backgrounds. The methodology of this technique involves the high-quality RNA extraction and preparation from specific tissues, followed by RNA amplification to facilitate hybridization, then the mRNA is converted into cDNA. This cDNA is further fragmented and biotin-labeled followed by the addition of the fluorescent molecule that binds to the biotin. Then, the hybridization is carried out, and the time required to complete the hybridization process signifies the sample concentration. Finally, the hybridized microarray is rinsed for removing the unbound chains. This is followed by microarray scanning where tagged fluorescent light detection indicates specific sequence hybridization at a specific point. The reading is performed by utilizing a laser, and the fluorescence emission is recorded by scanning. The fluorescence intensity determines the amount of probe bound to each sample (**Figure 4**).

However, there are several limitations associated with the microarrays: such as the expression levels detection is limited, and it is ineffective for extremely high and low expressive genes. This is dependent on prior existing knowledge, and sometimes it proves to provide error-prone outcomes. Additionally, the cross-hybridization between similar sequences leads to a reduction in appropriate detection. Hence, the results obtained from microarray need cross-validation by qRT-PCR, Northern blot, etc., by using appropriate reference genes [32].

## 6. RNA sequencing: the next-generation sequencing

RNA sequencing (RNA-seq) refers to quantifying the transcriptome using high-throughput sequencing methodology and computational methods [7]. The transcriptome is the set of various types of ribonucleic acid that are present in the cell such as messenger ribonucleic acid (mRNA), transfer ribonucleic acid (tRNA), ribosomal ribonucleic acid (rRNA), small nuclear ribonucleic acid (snRNA), non-coding ribonucleic acids (ncRNA), and others [33, 34]. The RNA-seq workflow includes total RNA extraction from a tissue sample, enrichment of RNA using either oligo (dT) or rRNA depletion, fragmentation of RNA (100–500bp), cDNA synthesis, and preparation of library then sequenced using various high-throughput sequencing methods, resulting into the short sequences from one end (single-end sequencing), which is faster and cost-effective than paired-end sequencing and also appropriate for quantification of gene expression levels. However, both ends (pair-end sequencing) generate more robust alignments and/or assemblies, which is found to be beneficial for gene annotation and transcript isoform discovery [7, 35]. The nucleotide sequences generate in a range between 30 bp and over 10,000 bp, vary with the sequencing method used [6]. Further, to study the expression level and transcriptional structure for each gene, the resulted sequences are aligned with reference genome sequences, available in databases. RNA-Seq reveals the genes that are active at a particular time, growth during the stages, or during treatment, read counts are used to studying the relative expression level. There has been continuous improvement made in the sequencing technology to obtain the finest result. There has been always huge importance of DNA sequencing in biological research that is hard to overstate. This sequencing technology helps to reveal the fundamental difference between the organisms. The limitations of the first-generation Sanger sequencing developed by Frederick Sanger and colleagues were overcome in the second-generation sequencing (SGS); likewise, in the third generation sequencing (TGS). Over the years, there have been wide innovations in the sequencing protocol, also a great elevation has been in the automation that has increased the capabilities of the DNA sequencing technology. Along with the technological advancements, it has made to be cost-effective that has resulted in the increased application and allows the parallel massive read of DNA of about hundreds of base pairs in a single run. The sequencing technology has shifted the researchers from computer to high-end servers, from code to programs, from single to multiple time points, and from single to multiple databases. **Table 2** provides the comparative account of the sequencing technologies.

The first-generation sequencing earlier involved gene fragmenting, cloning, and has a cumbersome manual analysis process. However, later it utilizes capillary gel electrophoresis, which involves the automation of capillary with polymers and sample loading and the computer-based detection of sequence [36]. This generates reads slightly less than 1 kilobase (kb) in length with an error rate of 0.001%. The second generation is also known as next-generation DNA sequencing (NGS) procedures that involve PCR-based *in vitro* cloning unlike the *in vitro* cloning in the first-generation sequencers [37, 38]. While the TGS that is available on the commercial scale doesn't involve cloning and can sequence a single DNA molecule [39]. Nevertheless, the Sanger sequencing platform has wide application as a gap-filling technology between contigs generated using NGS and TGS platforms.

The NGS involves a platform that can perform massively parallel sequencing of hundreds of thousands to hundreds of millions of different DNA fragments [40] with less template preparation. The NGS includes 454 pyrosequencing (Roche),

| Method | Commercial Released year | Typical Read length | Single Read Accuracy (%) | Reads per run | Time per run | Advantages | Limitations |
|---|---|---|---|---|---|---|---|
| Pyrosequencing 454 Life Sciences | 2005 | 700 bp | 99.9 | 1 million | 24 hrs | Long read size Fast | Runs are expensive Homopolymer errors |
| Sequencing by synthesis Illumina | 2006 | 50–600 bp | 99.9 | 1 million to 2.5 billion | 1–11 days | High sequence yield | Expensive equipment Requires high DNA concentrations |
| Sequencing by ligation SOLiD Sequencing | 2008 | 50 bp | 99.9 | 1.2 to 1.4 billion | 1–2 weeks | Low cost per base | Slower Issues with palindromic sequences |
| Combinatorial probe anchor synthesis cPAS-BGI/MGI | 2009 | 35–300 bp | 99.9 | 50 to 1300M per flow cell | 1–9 days | - | — |
| Ion semiconductor Ion Torrent sequencing | 2010 | 600 bp | 99.60 | Up to 80 million | 2 hrs | Less expensive equipment. Fast | Homopolymer errors |
| Nanopore Sequencing (Oxford Nanopore Technologies Ltd.) | 2011 | 500 kb | 92–97 | Up to 500 kb | 1 min–48 hrs | Portable | Lower throughput Lower accuracy |
| Single-molecule real-time sequencing Pacific Biosciences | 2011 | >100, 000 bases | 87 | 500,000 per Sequel SMRT cell 10 to 20 gigabases | 30 mins–20 hrs | Fast | Expensive |

**Table 2.**
*Comparative analysis of next-generation sequencing technologies for gene expression.*

Solexa sequencing (Illumina), ion semiconductor sequencing or Ion Torrent Proton sequencing, sequencing by oligonucleotide ligation and detection (SOLiD) system from Applied Biosystems massively parallel signature sequencing (MPSS). Among these, the former three are based on the principle of sequencing by synthesis while the SOLiD and MPSS employ the principle of oligonucleotide-template hybridization followed by ligation to the growing chain [41]. The MPSS is best suited for gene expression studies and utilizes enzymatic cleavage and ligation. This helps in distinguishing and quantifying the sample RNA from different species. The NGS technology has been used majorly for mRNA expression profiling, targeted resequencing, and biomarker discovery. It carries out the deduction of bases based on light color and intensity signals.

On the commercial scale, the short-read NGS sequencers available in the market are the short-read sequencers that possess sequencing ability of up to 600 bases, for example, Illumina, NovaSeq, HiSeq, NextSeq, MiSeq, Thermo Fisher's Ion Torrent sequencers BGI's MGISEQ, and BGISEQ. However, the NGS methods hold some limitations such as the short read length as a destructive effect of lasers on DNA and enzymes. Also, repetitive washing after each cycle affects the amount of DNA to be made available for sequencing. And as the plant genomes contain extensive repeat sequences, these short reads make the assembly of the genome sequences complicated. In addition, the heterozygosity and high/low GC-content regions could not be precisely assembled by utilizing the NGS. The NGS technology uses PCR for generating multiple copies of DNA fragments, which leads to biasness, and there is no uniformity in the quality of coverage of different genomic regions. This method relies on the principle of hybridization that requires a template as PCR generated millions of copies of a single DNA fragment, and as a result, the reaction does not occur in synchrony. Finally, in the case of NG sequencers, these asynchronous reactions ultimately lead to an increase in the error rate in the base sequence of the given fragment, which builds up through the cycles. However, the NGS platforms provide the software packages for "base calling" to minimize the error rate, and in addition, there are several base-calling algorithms present that reduce the error rate by ~5–30%.

Another limitation associated with the NGS technology is the time (several days) required for sample preparation despite generating the sequence data at a comparatively lower cost per base sequenced, the equipment, costs, chemicals, data storage, analysis, management, and other consumables increase the amount. The contrary, to the above limitations, the NGS technology still rules the commercial sector due to its capability of generating a huge amount of data with low per nucleotide cost. However, to deal with the above limitations in a successful way, the third-generation sequencing technology has been introduced.

## 6.1 Third-generation sequencing

As discussed above, the NGS sequencers are faster, cheaper, user-friendly with extremely high throughput. The TGS holds versatility and can successfully carry out several distinct analyses with much higher throughput and in a more cost-effective way than the NGS sequencers. Additionally, the TGS technologies only require a sequence of single DNA molecules. Hence, they do not depend on the in vivo cloning and PCR amplification and are extremely time-saving as they complete necessary template preparation in a few hours [39]. Therefore, they are often known as single-molecule sequencing (SMS) methods.

The TGS technology makes use of the enzymes DNA polymerase, fluorescence energy transfer, transmission electron microscopy, nanopores, and electronic detection. The TGS platforms are the long-read sequencers that produce reads of 10–15 kb. In the present scenario, Pacific Bio sciences' (PacBio) single-molecule real-time (SMRT) sequencing and Oxford Nanopore Technologies (ONT) nanopore sequencing are widely deployed [42]. However, the error rate of the TGS methods is reported to be quite higher about 10–15% as it makes use of single molecule so the error removal opportunity is less in comparison to the NGS, which carries out multiple copies of each fragment. Hence, this necessitates the requirement of the supporting technology to carry out the correction before and after the assembly process. Moreover, the supporting technology imparts support to existing genome assembly such as optical mapping method (Bio-nano), linked-read technology (10X genomics chromium system), and genome-folding-based technique HiC.

The selection of the appropriate sequencing technology has to be carried out based on a number of several implications, read coverage, accuracy, type of samples, DNA quality, quantity, and computation resources. However, in some cases, the combination of long and short-read sequencing platforms can be deployed as a better option for downstream analysis. This combination overcomes the individual limitations of both technologies and provides improved quality of whole-genome assembly.

In comparison with microarray, RNA-seq measures both low- and high-abundance RNAs, and it requires very little starting material, i.e., as little as 50pg, which made possible transcriptome studies of single cell over the tissue samples and helps in finer examination of cellular structures, expression level at a single-cell level along with an alternative transcript, novel transcript, and fusion genes. Several modifications of this RNA-Seq have been used for the identification of the candidate non-coding RNAs in plant species. A few of them are briefly described below.

### 6.2 Strand-specific RNA-Seq

Transcription of sense strand generates antisense transcript involved in the production of non-coding RNAs that are complementary with associated sense transcript. Antisense transcription was reported in nucleosomal-free regions such as promoters of bacteria, fungi, protozoa, plants, invertebrates, and mammals to carry out important regulatory functions. To identify the function and presence of antisense non-coding strand strands, there is a need for strand-specific RNA-Seq. Prevalent RNA-Seq does not preserve the information of sequenced transcripts. Beyond strand information, reads can be aligned to gene locus, but it will not give an idea about the transcription direction of a gene. Strand-specific RNA-Seq (ssRNA-Seq) helps to identify the transcribing genes, which overlap in various directions, and prediction of bouncing genes in organisms [43]. In ssRNA-Seq, the identity of the strand of DNA (sense or antisense) is preserved. This technique is also used to reveal the significant information of originating strand, a distinction between antisense and other non-canonical RNAs, which will be then used for enhancing the detection of a transcript from a sequencing experiment. For example, to uncover the sense and antisense transcript, mark off the boundaries of neighboring genes transcribed from both strands and study both non-coding and coding transcripts session level. A commonly used method for ssRNA-Seq is the dUTP [43], which involves the replacement of thymine nucleotide with uracil in

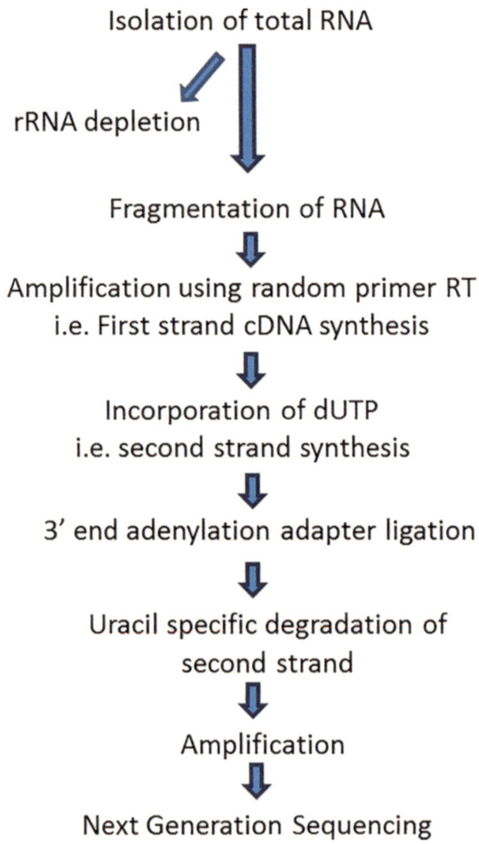


**Figure 5.**
*Flowchart for strand-specific-Seq.*

the complementary strand generated during second-strand cDNA synthesis. The complementary strands were further degraded by Uracil DNA Glycosylase (UDG); consequently, only the original strand remains back. Hence, in this way, the original strand used in the transcription can be identified, by aligning the sequence with the reference genome. By using strand-specific RNA-Seq, various novel lncRNAs have been identified in many plant species. One such example is reported in *Arabidopsis*, there has been the identification of a substantial amount of antisense transcription and long non-coding natural antisense transcripts (lncNATs) using this method. This cutting edge capable to give important information surrounding the transcriptome is a key to a greater understanding of the transcriptome. The methodology has been depicted in **Figure 5**.

### 6.3 RNA immune precipitation–sequencing (RIP-Seq) and CLIP-Seq

RIP-Seq refers to high-throughput sequencing of the interacting RNA, which is confined through immunoprecipitation of target proteins that helps to infer the mechanism of the posttranscriptional regulatory network. RIP-Seq maps the protein binding sites on RNA and produces RNA-protein complexes. Various long non-coding RNAs (lncRNAs) have been reported to date, while the functions of

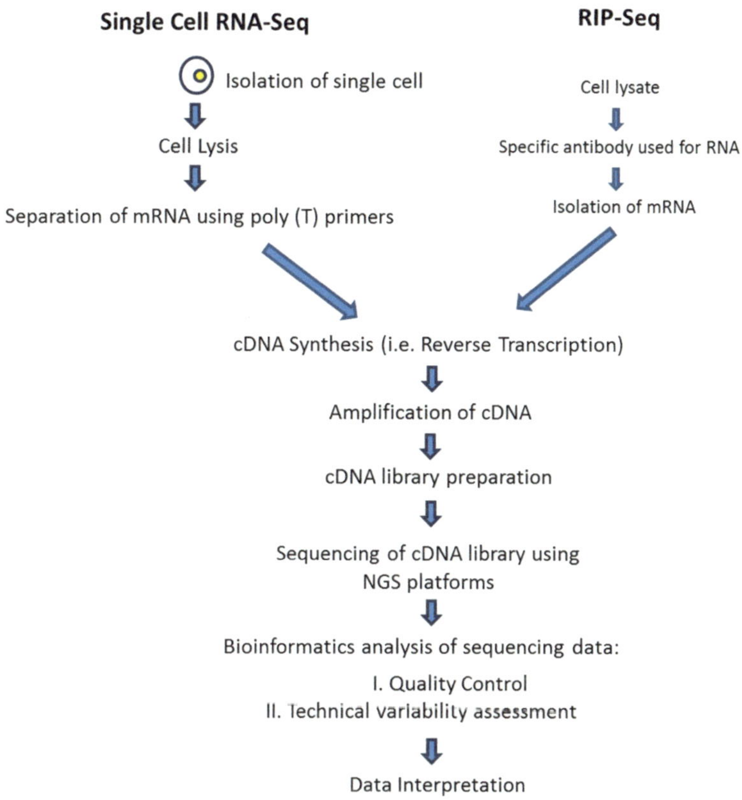


**Figure 6.**
*Flowchart of single-cell RNA-seq and RIP-Seq.*

many are still unclear. Hence, to reveal the significance of the lncRNAs, scientists have developed various technologies to study the RNA-RBP (RNA binding protein) interaction that is a critical mechanism regulating the translation. Mainly, there are two methods to characterize the functions of lncRNA, namely RNA immunoprecipitation sequencing (RIP-Seq) and cross-linking-immunoprecipitation sequencing (CLIP-Seq) (**Figures 6** and 7).

In RIP-Seq, proteins were used as bait to pull down the RNA from the sample, then protein targeted antibody is used for the immunoprecipitation of RNA-protein complexes, which are further purified under the optimum physiological condition to retain the native interactions. Followed by the RNase digestion, the extraction of the RNA protected by protein binding is carried out and is then reverse-transcribed to cDNA. Further, high-throughput sequencing has been carried out, and data analysis reveals the transcriptome-wide view of the protein-RNA/lncRNA regulatory network. Likewise, in CLIP-Seq, covalent binding between RNA molecules and RBPs under ultraviolet irradiation results in the improved binging strength of RNA binding proteins and their corresponding RNA targets.

## 6.4 Single-cell RNA-Seq

Rapid advancement in NGS-based technologies for genomics, transcriptomics, and epigenomics facilitated scientists to focus on individual cell characterization,

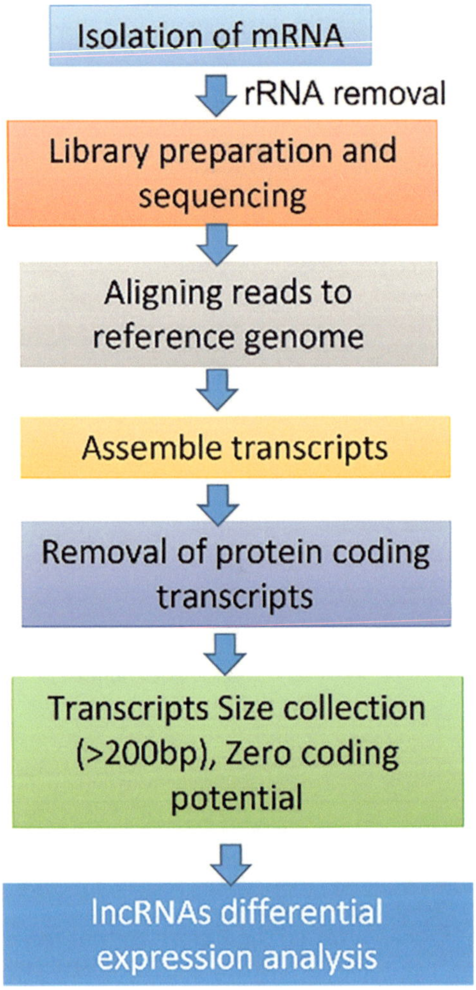


**Figure 7.**
*Flowchart of lncRNA sequenicng.*

which reveals significant novel and potentially expected discoveries. Studies on any biological system were carried out at the level of the organism, organ, or tissue. Nonetheless, cells of identical genotypes also change in the activity of only a subset of genes. Moreover, for a better understanding of a biological phenomenon, there is an essential requirement of a more precise transcriptomics study for individual cells that will further elucidate their role in numerous cellular functions. This will ultimately lead to a better understanding of gene expression in promoting beneficial and harmful states. There are six methods for sRNA seq includes, cell expression by linear amplification and sequencing (CEL-seq), droplet sequencing (Drop-seq), massively parallel single-cell RNA sequencing (MARS-seq), single-cell RNA barcoding and sequencing (SCRB-seq), switch mechanism at the 5' end of RNA template (Smart-seq, and Smart-seq2). In various plants such as *Zea mays*, *A. thaliana, Medicago truncatula*, rice, and *Glycine max*, expression pattern of genes was studied with the help of single-cell RNA seq methods [44]. However, there have been no reports about the utilization of the scRNA-seq to study plant lncRNAs. The major cause behind this is low sequencing coverage and the inability to capture and sequence non-poly A RNA. Although cell and tissue-specific roles and functional identification of lncRNAs in plants could be deduced using the scRNA-seq. In fluorescence microscopy, only a few genes can be studied under the response of cells to a specific signal or environment, while RNA-seq has been used for the study of differential gene expression levels with transcriptional differences of both coding and non-coding RNAs on a genome-wide scale. Single-cell transcriptomics has also been useful for the reconstitution of temporal transcription networks during developmental processes [45] or during the exposure of cells to external stimuli, all of which can be masked on a population level. In the below section, critical points for consideration are given.

## 7. RNA-Seq data analysis

### 7.1 Data quality control and reads mapping

Once RNA sequencing has been completed, the data generated need to be checked regarding the total numbers of reads generated, quality, and other requirements for sequencing. To remove the low-quality reads and base calls, filtering of reads and trimming of bases have been carried out, which are dependent on QC reports performed using RNA-SeQC [46] and RSeQC [47]. Reads of RNA-seq mapping with reference genome are quite more challenging than the procedure of mapping the general reads. This is mainly due to the synthesis of mRNA from the transcription process where the splicing out of introns and joining of exons in the gene make the RNA-Seq reads discontinuous (**Table 3**).

There are two approaches for mapping the RNA-seq: one is to construct a database of reference transcript sequences, which consists of currently annotated exons generated using a reference genome Reference transcript database is used to map such as BWA and Bowtie. Various examples of SpliceSeq [48], SAMMate [49], PASTA [50], RNASEQR [51]. While the other method detects *ab initio* splice junctions and is independent of genome annotation. Examples of *ab-initio* spliced mappers are TopHat/TopHat2 [52], MapSplice [53], HMMSplicer [54], GSNAP [55], MapNext [56], and STAR [57].

Recently, Salmon (https://salmon.readthedocs.io/en/latest/salmon.html#using-salmon), Sailfish (https://www.cs.cmu.edu/~ckingsf/software /sailfish/ ) and Kallisto (http://pachterlab.github.io/kallisto/) are being deployed.

The percentage of mapped reads varies with the different factors such as aligning methods and species although it is an important QC parameter (**Tables 4** and **5**). Additionally, several other critical factors need to consider such as rRNA reads and duplicate reads, which vary due to biological factors such as overrepresentation of a small number of highly expressed genes, or technical factor-like PCR over amplification. The RNA-Seq QC tools have good genomic coverage; it reports a percentage of reads often on the intragenic region (within genes including exons or introns) or intergenic regions (between genes). However, if a sequenced reference genome is not present to map the reads of RNA-Seq, then there exist two ways of

| | | |
|---|---|---|
| 1 | SAMStat | A tool evaluates unmapped, poorly and accurately mapped sequences independently to infer possible causes of poor mapping. |
| 2 | FastQC | A quality control tool for high-throughput sequence data. |
| 3 | RNA-SeQC | A tool with application in experiment design, process optimization and quality control before computational analysis. Provides three types of quality control: read counts, coverage, and expression correlation. |
| 4 | RSeQC | Analyzes diverse aspects of RNA-Seq experiments: sequence quality, sequencing depth, strand specificity, GC bias, read distribution over the genome structure and coverage uniformity. The input can be SAM, BAM, FASTA, BED files or Chromosome size file (two-column, plain text file). |
| 5 | Kraken | A set of tools for quality control and analysis of high-throughput sequence data. |
| 6 | dupRadar | An R package provides functions for plotting and analyzing the duplication rates dependent on the expression levels. |
| 7 | HTSeq | The Python script htseq-qa takes a file with sequencing reads (either raw or aligned reads) and produces a PDF file with useful plots to assess the technical quality of a run. |
| 8 | MultiQ | Aggregate and visualize results from numerous tools *(FastQC, HTSeq, RSeQC, Tophat, STAR, others.)* across all samples into a single report. |

**Table 3.**
*Tools for quality control of the transcriptome data.*

| | | |
|---|---|---|
| 1 | Cutadapt | Removes adapter sequences from next-generation sequencing data (Illumina, SOLiD and 454). It is used especially when the read length of the sequencing machine is longer than the sequenced molecule, like the microRNA case. |
| 2 | PRINSEQ | Generates statistics of your sequence data for sequence length, GC content, quality scores, n-plicates, complexity, tag sequences, poly-A/T tails, and odds ratios. Filter the data, reformat and trim sequences. |
| 3 | SnoWhite | A pipeline designed to flexibly and aggressively clean sequence reads (gDNA or cDNA) prior to assembly. |
| 4 | AlienTrimmer | Implements a very fast approach (based on *k*-mers) to trim low-quality base pairs and clip technical (*alien*) oligonucleotides from single- or paired-end sequencing reads in plain or gzip-compressed FASTQ files. |
| 5 | Trimmomatic | Performs trimming for Illumina platforms and works with FASTQ reads (single or pair-ended). Some of the tasks executed are: cut adapters, cut bases in optional positions based on quality thresholds, cut reads to a specific length, and convert quality scores to Phred-33/64. |

**Table 4.**
*Tools for trimming and adapter removal.*

analysis of RNA-Seq data. First, use a reference genome of related species to map reads, and another is to assemble the target transcriptome *de novo*. Many *de novo* transcriptome assemblers are available, which include Oases [58], SOAPdenovo-Trans [59], Trinity [60], and Trans-ABySS [61]. The point that has to be considered is reference genome of related species must have genome similarity ~85% or more with the species of study, otherwise better to go with the *de novo* assembly approach.

| | | |
|---|---|---|
| **Short (Unspliced) Aligners** | | |
| 1 | Subread | Expression analysis |
| 2 | Bowtie | A short aligner based on the Burrows–Wheeler transform algorithm and the FM-index. Bowtie tolerates a small number of mismatches. |
| 3 | Burrows–Wheeler Aligner (BWA) | A software package for mapping low-divergent sequences. |
| 4 | Bowtie2 | Aligns sequencing reads to long reference sequences that supports gapped, local, and paired-end alignment modes. |
| 5 | PerM | Genome-scale alignments for hundreds of millions of short reads produced by the ABI SOLiD and Illumina sequencing platforms. |
| 6 | ZOOM | Short aligner of Illumina/Solexa 1G platform, uses extended spaced seeds methodology building hash tables for the reads and tolerates mismatches and insertions and deletions. |
| **Spliced aligners** | | |
| 1 | RNA-MATE | Pipeline for alignment of data from Applied Biosystems SOLID system. |
| 2 | Erange | Alignment and data quantification to mammalian transcriptomes. |
| 3 | RUM | Alignment based on a pipeline, being able to manipulate reads with splice junctions, using Bowtie and Blat |
| 4 | RNASEQR | Tools used for alignment. |
| 5 | SAMMate | |
| 6 | SpliceSeq | |
| 7 | X-Mate | |
| *De novo* **splice aligners** | | |
| 1 | HiSAT | Alignment program for mapping RNA-seq reads. |
| 2 | HISAT2 | Alignment program for mapping next-generation sequencing reads. |
| 3 | HMM Splicer | Canonical and non-canonical splice junctions in short-reads. |
| 4 | GMAP | A Genomic Mapping and Alignment Program for mRNA and EST Sequences. |
| 5 | Pass | Aligns gapped, ungapped reads and also bisulfite sequencing data. |
| 6 | QPALMA | Predicts splice junctions supported on machine learning algorithms. In this case the training set is a set of spliced reads with quality information and already known alignments. |
| 7 | SuperSplat | Algorithm splits each read in all possible two-chunk combinations in an iterative way, and alignment is tried to each chunck. |
| 8 | SoapSplice | Tool for genome-wide ab initio detection of splice junction sites from RNA-Seq, a method using new generation sequencing technologies to sequence the messenger RNA. |
| 9 | RASER | Reads aligner for SNPs and editing sites of RNA. |
| *De novo* **splice aligners (also for annotation )** | | |
| 1 | STAR | Align long reads and can reach speeds of 45 million paired reads per hour per processor. |
| 2 | TopHat | Alignment of shotgun cDNA sequencing reads. |
| 3 | Subjunc | Uses all mappable regions in an RNA-seq read to discover exons and exon-exon junctions. |

**Table 5.**
*Tools for alignment of the transcriptome data.*

## 7.2 Data normalization, differential gene expression, and splicing variant analysis

Normalization refers to removing the technical bias and unwanted variation in the total read count of different samples, which helps to focus on sample difference. In RNA-Seq, genes that are highly expressed, i.e., transcribed, mean more reads will be present for the same gene. However, the critical factors that need to be considered while applying this basic principle are the sequencing depth and length of gene transcript. Comparing reads of different genes over the sample in different treatments helps to normalize the number of reads for each gene (**Table 6**).

Reads per kilobase per million (RPKM) and fragments per kilobase per million mapped reads (FPKM) are the two simplest early normalization approaches in RNA-Seq data; nevertheless, additional tools such as DESeq and edgeR are also commonly used for normalization. In FPKM, the gene expression is normalized during software such as StringTie, which helps in transcript assembly and RNA-seq quantification. Then FPKM value is generated for the gene expression, where the higher value indicates the increased gene expression. Moreover, this software has also been utilized for the identification of the alternative transcripts generated during the splicing of mRNAs during developmental stages. The transcripts per million (TPM) based on depth-normalized counts and counts per million reads mapped (CPM) based on length-normalized are also used as metrics depending upon the experimental consideration.

To identify the differentially expressed genes, various models are available such as bayseq [62], Cuffdiff / Cuffdiff2 [45], DEGSeq [63], DESeq/DESeq2 [64], edgeR [65],

| | | |
|---|---|---|
| 1 | BaySeq | It is a Bioconductor package to identify differential expression using next-generation sequencing data, via empirical Bayesian methods. |
| 2 | DESeq | It is a Bioconductor package to perform differential gene expression analysis based on the negative binomial distribution. |
| 3 | Derfinder | It helps to annotation-agnostic differential expression analysis of RNA-seq data at base-pair resolution via the DER Finder approach. |
| 4 | DiffSplice | It is a method for differential expression detection and visualization, not dependent on gene annotations. |
| 5 | EdgeR | It is an R package for analysis of differential expression of data from DNA sequencing methods, like RNA-Seq, SAGE or ChIP-Seq data. |
| 6 | EdgeRun | It is an R package for sensitive, functionally relevant differential expression discovery using an unconditional exact test. |
| 7 | MetaDiff | Differential isoform expression analysis using random-effects meta-regression. |
| 8 | MMSEQ | It helps to estimating isoform expression and allelic imbalance in diploid organisms based on RNA-Seq. |
| 9 | Rcount | It is a simple and flexible RNA-Seq read counting. |
| 10 | rDiff | It is a tool that can detect differential RNA processing (e.g. alternative splicing, polyadenylation or ribosome occupancy). |
| 11 | StringTie | It is an assembler of RNA-Seq alignments into potential transcripts. |
| 12 | TIGAR | Transcript isoform abundance estimation method with gapped alignment of RNA-Seq data by variational Bayesian inference. |
| 13 | TimeSeq | It helps to detects differentially expressed genes in time course RNA-Seq Data. |

**Table 6.**
*Tools for quantitative analysis and differential expression.*

Limma Voom (https://www.bioconductor.org/packages/devel/bioc/vignettes/limma/) While reads counting software T-Seq are used for the counting aligned reads for overlap of reads and edgeR or Deseq2 is used to find out differentially expressed genes in the form of heat map, which shows higher expression by dark-colored pattern and decreased expression denoted by pale color relative to controls [65]. Sequencing depth (number of times a sample is sequenced) and high coverage (number of reads) obtained after sequencing are the key important factors to uncover the low-level expressed novel transcripts such as lncRNAs. Deep RNA-Seq helps to identify the novel lncRNAs in plants by resequencing cDNA fragments.

Similarly, SpliceSeq quantifies and compares reads covering exons, while the splicing junction approach is used to identify the change in splicing pattern. Many of these methods commonly focus on the level of splicing events instead of full-length splicing variants. Various methods are available such as MISA [66], ALEXA-Seq [67], FDM [68], rDiff [69], and rSeqDiff [70]. Genome-independent methods for the analysis of splicing variants, especially used in case of species not having sequenced reference genome or huge variation in RNA transcript (diseased condition) compared with the reference genome. To assemble and differentiate splicing variants, methods are based on transcriptome preassembled from RNA-Seq reads, i.e., transcriptome-based approach includes RSEM [71], IsoEM [72], BitSeq [73], and recently developed such as Rnnotator [74] and KisSplice [75].

### 7.3 Functional analysis of identified genes

Once the differentially expressed genes are revealed, there comes the need to understand the biological functions of those genes. Functional analysis of identified genes is an important part of data analysis, and it has been carried out at multiple levels such as biological pathways, gene ontology, and gene networks. Many different tools are available for functional analysis such as DAVID, g:profiler and clusterProfiler [76, 77] used for the analysis of GO and biological terms, GSEA [78], which

| S. no. | Tools | Remarks |
|---|---|---|
| 1 | Tombo | A suite of tools for the identification of modified nucleotides, analysis and visualization of raw nanopore signal from nanopore sequencing data. |
| 2 | IDP | Tool for *de novo* transcriptome assembly and isoform annotation by hybrid sequencing. |
| 3 | NanoMod | Detection of DNA modifications using Nanopore long-read sequencing data. |
| 4 | Pinfish | Pinfish is a collection of tools helping to make sense of long transcriptomics data (long cDNA reads, direct RNA reads). |
| 5 | TAPIS | TAPIS (Transriptome Analysis Pipeline from Isoform Sequencing) is a program for correcting and aligning long reads with/without the second generation reads, transcript clustering, novel and full-length splice isoform detection, and identification and analysis of polyadenylation (poly(A)) and alternative poly(A) (APA). |
| 6 | SQANTI | SQANTI provides a wide range of descriptors of transcript quality and generates a graphical report to aid in the interpretation of the sequencing results. |
| 7 | Tama | software was designed for processing Iso-Seq data and other long read transcriptome data. |

**Table 7.**
*Tools for annotation.*

| | | |
|---|---|---|
| 1 | BamView | BamView is a free interactive display of read alignments in BAM data files |
| 2 | IGV | The Integrative Genomics Viewer (IGV) is a high-performance, easy-to-use, interactive tool for the visual exploration of genomic data. |
| 3 | BrowserGenome | Web-based RNA-seq data analysis and visualization. |
| 4 | ABrowse | A customizable next-generation genome browser framework. |
| 5 | Savant | Savant is a next-generation genome browser designed for the latest generation of genome data |
| 6 | EagleView | EagleView is an information-rich genome assembler viewer with data integration capability. |
| 7 | TBro | It is a transcriptome browser for de novo RNA-sequencing experiments. |
| 8 | MicroScope | It is a comprehensive genome analysis software suite for gene expression heatmaps. |
| 9 | MatchAnnot | MatchAnnot is a python script which accepts a SAM file of IsoSeq transcripts aligned to a genomic reference and matches them to an annotation database in GTF format. |
| 10 | Iso-Seq | The Iso-Seq method produces full-length transcripts using Single Molecule, Real-Time (SMRT) Sequencing. |
| 11 | IsoSeq-Browser | Interactive visual analytics tool for long-read RNA sequencing (Pacific Biosciences' isoform sequencing (Iso-Seq) techniques). |

**Table 8.**
*Tools for data visualization of transcriptome data.*

is used for functional analysis of the entire gene set, and IPA (ingenuity pathway analysis) for gene network analysis. The online sources for gene annotation such as OmicsBox (https://www.biobam.com/omicsbox/), Panzzer2 (http://ekhidna2.biocenter.helsinki.fi/sanspanz/), EggNOG (http://eggnog-mapper.embl.de.) are widely deployed (**Table 7**).

### 7.4 Data visualization

After annotation, variants can be visualized using genome browsers and visualization tools. Many RNA-Seq data visualization tools are available such as Genome Browser (https://genome.ucsc.edu/), Integrated Genome Viewer (https://software.broadinstitute.org/software/igv/), and Jbrowse (https://jbrowse.org/jb2/). Alternative splicing visualization tools such as Alexa-Seq, SpliceSeq, SpliceGrapher, and SpliceViewer are also available. These visualization tools help to understand the information of variants including reads, mapping reads, and annotation information such as consequences, scores, and impact of variants. To demonstrate the large changes in gene expression volcano plot can be used. In the volcano plot, each dot is a representation of a gene, whereas the x-axis and y-axis represent the log-fold change based on FPKM values and log10 (p-values), respectively (**Table 8**).

## 8. Analyses with PacBio and NanoPore datasets

Long read sequencing of the transcriptome is done generally to qualitatively understand the expression of the genes/transcripts in the organism. This is done by understanding where the genes are localized and whether there are events such as fusions/deletions impacting the genes. Unlike Illumina or other short-read sequencing approaches, where the number of reads can be in millions, thus capturing the expression multiple times, long reads generally are in a few thousand but offer the capture of full-length transcripts depending on the libraries prepared. For starters, earlier in 2010–14, most of the expression data used in the range of 25–100bp long were either paired or unpaired. Currently, the short-read technology can sequence up to 250bp long, which essentially means that the transcripts are going to be fragmented a few times.

PacBio's IsoSeq and Nanopore's direct cDNA/amplified cDNA sequencing can capture the complete expressed transcripts in the range of up to 90Kbp, with the median being around 1400 for PacBio and 770bp for Nanopore (depending on Nanopore's preps). Generally, IsoSeq and direct cDNA capture are done to confirm the existence of long repetitive regions, gene isoforms, and gene fusions. This then aids in annotation of the genome, capturing alternative splice-sites, etc. **Figure 8** depicts the analysis to be executed in the transcriptomics datasets.

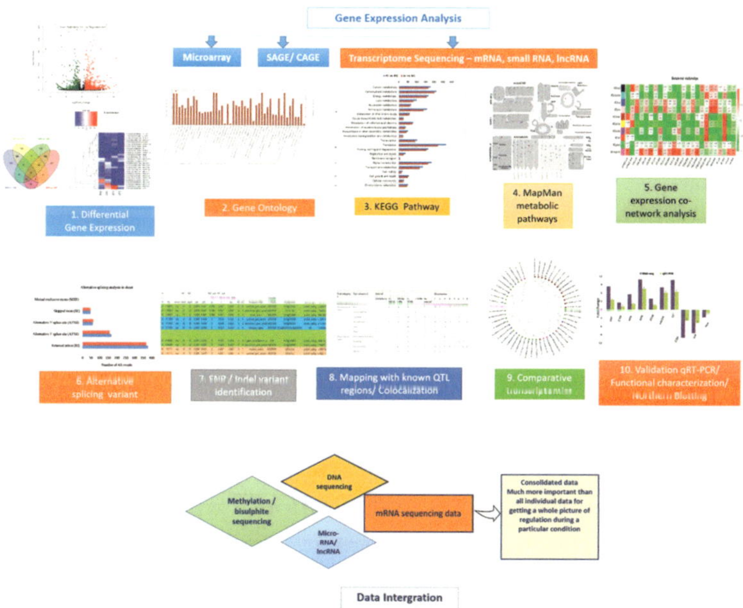


**Figure 8.**
*A complete overview of the downstream analysis to be executed for the transcriptomics datasets.*

### 8.1 Generalized workflow for PacBio/Nanopore

The general steps of analyses for IsoSeq are as follows:

1. Generate reads of the insert with multiple passes to ensure high-quality reads with Q>30.

2. Identification of the reads that represent full-length transcripts based on the presence of Poly-A tails.

3. Cluster the reads iteratively using the longest reads and polish the read to obtain high-quality consensus.

4. Map the consensus to the genome with long read aligners such as Minimap2 or GMAP.

5. Identify the genomic regions where the read maps to derive gene models based on identity or coverage thresholds.

The general steps of analyses for Nanopore are as follows:

1. Perform base-calling with Guppy using the model suitable for Minion, Gridion, or Promethion based on the flowcell used.

2. Identify reads with primers on both the ends, these will likely be the full-length transcripts.

3. Perform all vs all overlap with the reads to get overlaps with Minimap2 and do consensus calling with Racon.

4. Align the consensus reads to the genome to derive gene models using Minimap2 or GMAP.

Once the gene models or their transcript sequences are identified, the next few steps are to understand if these sequences exhibit any sort of a function. For example, are these sequences coding or non-coding, are these protein-coding transcripts, or are these long non-coding transcripts? Fortunately, there are multiple approaches to solve this issue.

One of the easiest ways to know this is to evaluate the coding potential of these transcripts. This can be done using tools such as Coding Potential Calculator, Coding–Non-Coding Index, and Coding Potential Assessment Tool. Once we know which transcripts are coding based on the results from these approaches, we can select the remaining ones and call them long non-coding RNAs.

Another approach is to convert the transcripts into peptide sequences using something like TransDecoder or Evidential Gene, which can then be used to get functional assignments from PFAM, RFAM, eggnog, InterPro databases, etc. The above approaches do not necessarily require the presence of a reference genome as recent developments/tools can directly use partial-order alignments (POAs)-based approaches to generate clusters, which can be then used to derive consensus directly.

## 8.2 Combinatorial analyses with short reads

The consensus sequences can then be used in place of a reference genome to study the transcriptome of the organism directly, by using short-read data to quantify the expression of the transcripts. The long reads provide a qualitative expression of the organism, whereas the short reads will give the actual measure of expression due to the sheer quantity of data. Once the gene models from long reads are available, annotated, and curated approaches such as Salmon/Kallisto/RSEM, etc., can be deployed to quantify the expression using short reads based on which differential expression can then be performed. Similarly, tools such as SQUANTI/TAMA, etc., exist, which can use short-read data to augment the long-read data by annotating with CAGE peaks, polyA sites, NMD prediction, etc., which can be used for downstream analyses (Appendix).

## 9. Transcriptomic databases

Transcriptomic studies provide enormous information beyond the aim of experiments, which can serve as a base for other scientific communities. This large amount of data generated through experiments may be deposited in publicly available databases. In transcriptome, it quantifies the expression of genes along with small RNA and noncoding RNAs in cells, organs, particular growth stages, or stress conditions [79, 80]. Identified information in transcriptomic studies such as differentially regulated genes under the stress condition that can be targeted in the crop improvement programs [80–82]. Presently, a lot of information is available publicly through databases for *in-silico* studies. NCBI GEO is one of the highest updated and curated databases, which provides information regarding microarray data, RNA sequences, and functional annotation [83]. In addition to the earlier discussion, there are many other databases available that account for the RNA co-expression (http://atted.jp), plant-pathogen interaction, phosphorylation sites, RNA editing events, and transcription factors.

## 10. Validation of the RNA-seq data

The gene expression data obtained from RNA-seq studies need to be validated experimentally. The high-throughput large datasets emanate a large number of genes, and practically validating all the relatively expressed genes has limitations. Hence, validation can be performed on small or large subsets as per the design, sampling, and tissues of the experiment. Such validation should be ideally done using the same samples subjected to RNA-seq or microarray. Quantitative real-time PCR (qRT-PCR) is the most widely used technique for the validation of gene expression on account of reliability, accuracy, and sensitivity. It is considered a medium-throughput gene expression analysis technology and is largely used for the validation of transcriptome studies. Other relatively less deployed techniques are translational fusion reporters using reporter genes, functional assays, etc. The virus-induced gene silencing (VIGS) is an RNA interference-based technology deployed to transiently knock down the target gene expression by utilizing modified plant viral genomes. It is an emerging resourceful tool for functional validation of more number of genes [84].

The qRT-PCR remains a widely adopted mandatory technique for the validation of gene expression. Nevertheless, it holds the best with the use of the same samples as assayed for RNA-seq. This one is always well meant when other replications from the same sampling population are assayed using the qRT-PCR . The reference genes or housekeeping genes or endogenous genes whose expression is expected to be stable in a particular tissue at a given time play a critical role in the quantification of gene expression. The variables in the experiment are taken care of by the appropriate usage of reference genes in the experiment [85]. The most commonly used reference genes are 18s rRNA, GAPDH, actin, ubiquitin, elongation factor, tubulin, etc. The selection of a reference gene set is very crucial in differential expression studies as it is known that varying reference genes work in a particular tissue in a spatiotemporal manner. The available software/tools for validation of the reference genes are delta (ΔCt), geNorm, qBASE, NormFinder, BestKeeper, and RefFinder [86]. The selection of the number of reference genes depends on the M value and V value.

Livak's $2^{-\Delta\Delta CT}$ method is the most popular method for quantifying relative gene expression using the target and reference gene Cq (quantitative cycle) or Ct (cycle threshold) used Minimum Information for Publication of Quantitative Real-Time PCR Experiments (MIQE) guidelines to describe the information required for publishing the qRT-PCR-related data in terms of transparency, accuracy, reliability. The readers are encouraged to refer to [87, 88].

### 10.1 Critical points for qRT-PCR validation

**A.** Designing qRT-PCR primers having desirable amplicon size (60–150 bp), GC content (40–60%), primer length (18–25 bases), high Tm (temperature melting 60–62°C), and without formation of dimers, loop structures. The primers can be checked using online tools for the secondary structure formation and thermodynamics parameters.

**B.** Selection of sample sets (as per the experimental design) for experimental validation and the number of samples for validation in replications.

**C.** Priori checks using semiquantitative PCR for cDNA template concentration, primer concentration, and PCR reaction components.

**D.** Selection of a validated set of reference/endogenous genes in the experiment.

**E.** Appropriate use of positive, negative controls, negative template controls in plate setup.

**F.** Number of biological replications (at least three) and technical replications (two) for statistical inference.

## 11. Critical factors to be considered in expression studies

Numerous points need to be taken into account for gene expression studies. First, and the most important, factor to be considered is the sample tissues. As the expression studies have an enormous variation depending on the developmental stage and the aim of the experiment. Hence, the time of sample collection and proper storage holds crucial importance while experimenting. The selected individuals should be representative of the species and should possess strong genetic background. This will enable to extract adequate information on a large scale. Further, the isolated nucleic acid quality and quantity should be thoroughly checked before performing the NGS to get accurate outcomes. Several points need to be taken into consideration for the selection of the required sequencing platform and assembly tools/software/program to get proper and accurate results. The choice of the sequencing platform to use will influence the cost and success of the assembly process. Different types of sequencing platforms generate different types of data that can be analyzed using different assembly programs. The assembly program is very specific for the type of data to be analyzed so the analysis pipeline should be decided before prefer sequencing. The biological factors include the selection of individuals with pure genetic backgrounds and good representatives of the species to be used for DNA isolation. The basic studies regarding the biochemical, morphological, and physiological should be known. When considering the technical factors, the computational tool has an enormous role as for the proper genome assembly, to run proper analysis process, storage. The accurate selection of the annotation program to be used is also a critical step. So that the gene/transcript has a low copy number and must not escape from the study. The proper stringency level of the bioinformatics tools/software should be maintained for the final interpretation of the data generated. For the proper alignment of assembly and annotation of coding regions, the RNA sequencing data must be generated by extracting RNA of the same sample used.

### 11.1 Critical points for biology consideration before the start of transcriptome sequencing

A. To understand gene expression at a given time in a tissue spatiotemporal

B. To know the changes at different time points happening in a tissue

C. To compare the changes in expression across tissues in varying organisms

D. To understand the biology of affecting/invading microorganisms in another plant/ animal

E. To pinpoint a specific gene/transcript/s responsible for that trait/condition

### 11.2 Critical parameters affecting a transcriptome

A. Quality of reads (Phred score)

B. Availability of the reference genome

C. The completeness of the transcript/gene

D. GC content

E. Number of transcripts in the assembly (assembly thinning)

F. The correctness of the transcripts

G. Replications for statistical analysis of data

H. Functional significance

I. Validation of transcripts

J. Phylogeny, etc.

### 11.3 Critical points before the start of the experiment

A. Know all about the biology of the sample/trait

B. Time points of sampling/data required at that particular need to be checked or validated

C. Need to cross-check and refer as per biological experiment under consideration

D. Critical time points at which data need to be recorded

E. Keep the data sorted based on numbers, maximize using excel/filter based on various parameters such as e value, FPKM, q value, p-value stats

F. Use more than two/three tools for each data

G. Keep the stringencies at different levels and observe the data/data distribution

H. Always check back the data with the available information from RNA-seq, ESTs, genome data, back references

I. Validation using qRT-PCR select the maximum number of set of transcript based on function and transcripts falling in one pathway for validation

### 11.4 Critical points for data accuracy

A. Cross-check the data at every step of data analysis

B. Using two tools/software will increase accuracy

C. Stringencies in terms of parameters and statistics should be checked at every step

D. Annotation can be checked with the known set of proteins of the nearest genome or with the RNA-seq data available in the public databases

E. The functionally relevant transcripts (relatively higher or lower) should be biologically validated

F. The gene expression validation using a particular significant pathway or metabolic process or a set of specific gene families will hold a higher confidence level for data validation

## 12. Integration of transcriptomics with other techniques for unravelling the gene expression

The techniques of gene expression especially the RNA-seq are widely deployed in plants, humans, and animal sciences for quantitative and qualitative profiling. The data emanated from the RNA-seq can be analyzed for differential gene expression, annotation, isoform identification, metabolic pathways, domain identification, alternative splicing variant identification, insertion-deletion variations, single nucleotide variants, gene expression co-network analysis, mapping with already identified regions, or quantitative trait loci (QTLs). Along with the mentioned downstream analysis, the RNA-seq gene expression data can be generated and integrated with other techniques for better understanding of specific human or animal processing depending on the tissue, complexity, and the metabolic and biological processes. The data can be generated *de novo* (sequencing from the samples), or the available data in the publically available databases can also be mined and utilized [89].

Specific to the human and animal sciences in view of the complexities associated with the varying diseases, responses to disease conditions and healthcare medications, the following advances of RNA-seq techniques have been majorly deployed for studying the gene expression. The spatial gene expression in tissue sections retains the precise location of biological molecules in tissue samples and then can be sequenced for knowing the morphological differences. Similarly, the formalin-fixed, paraffin-embedded (FFPE) tissues and antibodies tagged with cell-surface proteins can be sequenced. For better analysis of the relative gene expression studies, RNA-seq techniques are being combined with DNA methylation, degraded RNA samples, protein, and chromatin studies for a thorough understanding of gene expression at a given time point. The circulating RNA can also be captured by using modified protocols (during the initial isolation steps from tissues) and sequenced for the identification of transcripts. In the clinical aspects of the treatment of diseases, it is essential to characterize the immune repertoire at the single-cell level. The techniques such as cellular indexing of transcriptomes and epitopes by sequencing (CITE-Seq) combine single-cell RNA-Seq with cell surface protein analysis and facilitate analysis of cell-surface proteins. The specific region of interest can also be sequenced using an enrichment probe-based approach that can also be deployed to target the transcripts of interest called targeted enrichment RNA-seq. The understanding of alternative splice variants also forms a major application of RNA-seq in clinical research [90].

The RNA-bulk seq is a modified technique of bulk segregant analysis wherein the extreme bulks are made for the identification of QTLs and the gene expression patterns associated with the trait of interest. The RNA samples from contrasting types of tissue are bulked and sequenced called RNA-bulk sequencing, which can be combined with spatial RNA-seq for quantitating the gene expression of tissues at a given time [91].

The epi-transcriptomics pertains to the transcriptome analysis to understand the RNA modifications such as N6-methyladenosine, 5-methylcytidine, and 5-hydroxylmethylcytidine. Specific antibodies are used for precipitating the RNA (RNA immunoprecipitation (RIP)) with modifications that are then sequenced on a high-throughput platform. The Oxford Nanopore RNA-Seq can detect the modifications directly without the need for antibodies [92].

Dual-RNA seq is a persuasive method for analyzing the simultaneous gene expression patterns of the host and microorganism during their interaction. The interaction can be beneficial as has been observed in the growth-promoting microorganisms or during an infection process. The transcripts from the host and the microorganisms are concurrently captured, and the genome-wide transcriptional changes from the host as well as from the microorganism can be accessed. This technique unravels the mechanism of the beneficial organism or invading pathogen enabling the understanding of the effectors and molecular processes of host colonization. Nevertheless, the practical procedures of isolating the interactive transcriptome require specialized protocols and further bioinformatics analysis [93].

The RNA-seq datasets generated through sequencing can be utilized further for mapping the trait of interest. The genetic variants present in a particular region called expression quantitative trait loci (eQTLs) regulate the expression levels of local or distant genes and explain the variation in the gene expression. The genome-wide association studies (GWAS) results can be integrated with the eQTL data in an approach called transcriptome-wide association studies (TWAS). The gene expression levels for GWAS samples can be combined with the gene expression datasets for that trait expression (trait values) in order to identify the gene-trait associations (the involvement of that genic region/genes associated with the trait). The TWAS is a potential approach to ascertain the causal genes at the GWAS loci [94]. In addition, the differentially expressed genes can be mapped with the already known reported QTLs for a particular trait of interest. These co-localized genes increase the confidence of the study in terms of linkage with the QTL.

In the specific applications of understanding the biogenesis, development of non-coding RNA, transcription sites, or finding the binding sites of transcriptionally active RNA polymerase II (RNAPII), the Global Run-On sequencing (GRO-seq) is utilized. The GRO-seq allows the unbiased mapping of nascent transcripts. Brominated nucleotides (5-bromouridine 5′-triphosphate (Br-UTP)) are deployed for immunoprecipitation and enrichment of nascent RNA followed by cDNA conversion and sequencing [95].

## Appendix Example: bioinformatics pipeline of transcriptome analysis

1. QC of the data—FastQC
2. Adapter low-quality data trimming—FastP
3. Alignment—STAR/HiSAT2
4. Alignment QC—Biotype plot, duplicate marking, strandedness identification
5. Quantification—Feature counts
6. Sample correlation and principal components—DESeq2
7. Comparative analyses—DESeq 2: Volcano plots, HeatMaps
8. Term enrichments—GO overrepresentation, reactome, and KEGG pathways

### 1. QC of the Data—FastQC

FastQC is generally used to judge the quality of the data based on Phred Scores. Phred Scores are negative log score that is used to assign the quality of the base that is called.

| Phred Quality Score | Probability of incorrect base call | Base call accuracy |
|---|---|---|
| 10 | 1 in 10 | 90% |
| 20 | 1 in 100 | 99% |
| 30 | 1 in 1000 | 99.9% |
| 40 | 1 in 10,000 | 99.99% |
| 50 | 1 in 100,000 | 99.999% |
| 60 | 1 in 1,000,000 | 99.9999% |

**Q = -10 log10P. The probability of the base calling increases, quality increases**

FastQC gives visual confirmation that all is well with the data. GC content plots can be used as an assessment check to know if there are any contaminants present in the sequences. Weird distribution at the starting or the ending of the reads can be a signature of library artifacts or systemic biases of the sequencer, esp. Illumina.

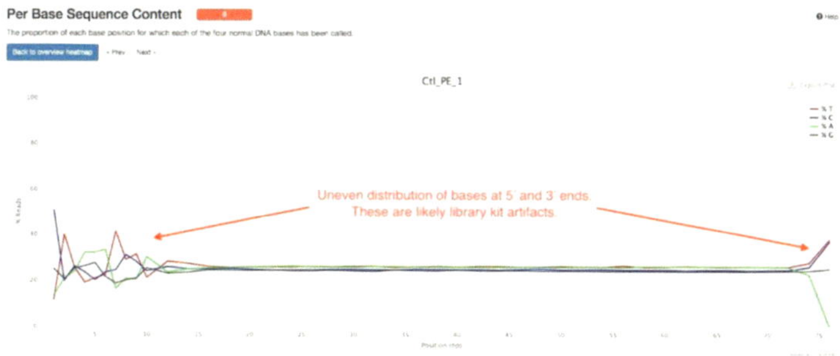


Distribution of the bases at the starting and ending of the reads is an example of biases of the sequencer of improper library preps

**2. Adapter/low-quality data trimming—FastP**

Fast is a brilliant all-in-oneQC tool. It gives a summary of the data before the removal of bad regions in a read and after removing the bad regions.

Bad regions can be specified as a stretch of bases with a lower quality than expected. For example, if the average read quality is to be Q35, i.e., roughly greater 99 than 95% accuracy, but a few bases have a Q15 score, i.e., around 93% accurate, then we can remove these.

FastP can also trim off the head or tail of the reads in **Figure 2** above, do a duplicate analysis to see how much data are duplicated, and what is the common motifs present in the data.

**3. Alignment—STAR/HiSAT2**

Fast trimmed reads are aligned to the genome using STAR/HiSAT2, which are both splice-aware aligners. This means for higher organisms such as eukaryotes where the mRNA is formed by splicing out the introns, the aligner can try to truncate the reads partially at the exon-intron boundary and try to align starting at the next intron-exon junction.

STAR uses a sparse function to store the representation of the genome, whereas HiSAT2 stores the indices in a hierarchical linked manner to align the reads.

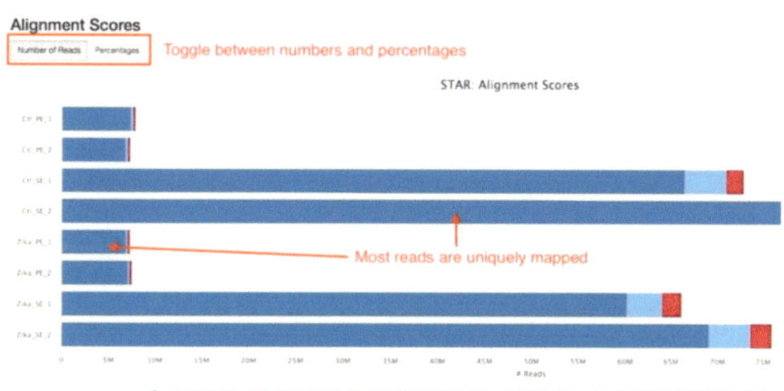


STAR alignment scores

A good sample would have the largest set of aligned reads mapped uniquely. A large representation of multi-mapped reads suggests rRNA contamination

**4. Alignment QC—Biotype plot, duplicate reads, strandedness identification**

The alignments can specify a plethora of information about the samples and the sequenced data. A few pointers to note are:

a. In the case of RNASeq, the majority of the reads should be aligned to the exons.

b. The major biotype should be related to protein-coding genes.

c. Duplicate reads do arise in RNASeq, and you can either mark them as duplicates or not.

d. Strandedness of the data. Some genes have their regulatory bodies on the opposing strand. Specifying to the provider to perform stranded sequencing can help identify if the gene is getting expressed or depressed due to the regulatory effect.

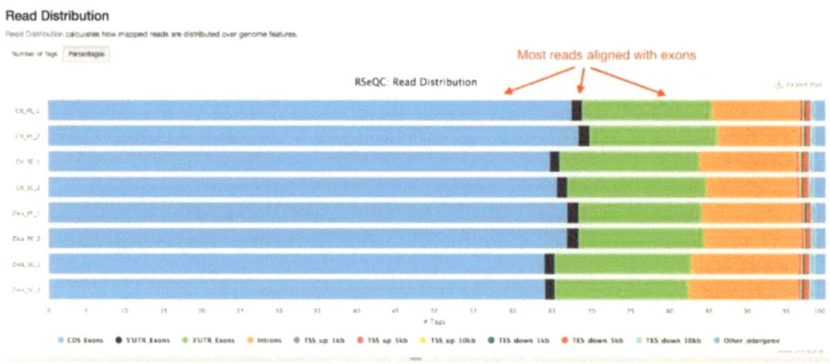


Sequencing reaction type also affects the duplicacy rates

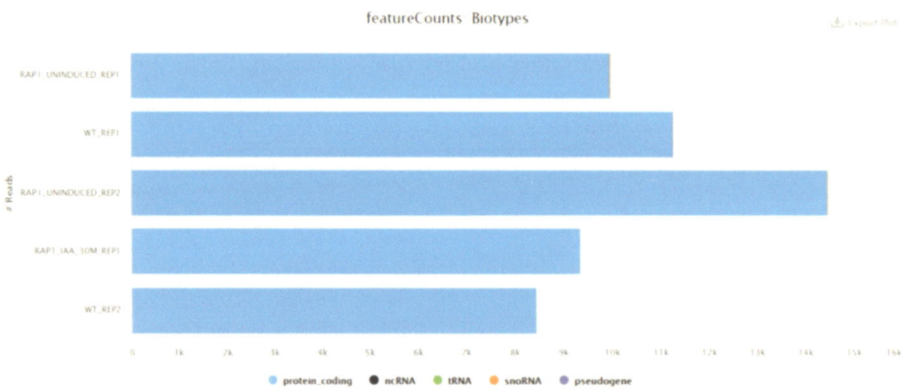


Identified by annotating the alignments, can tell if the library prepared has captured the protein-coding genes or auxillary contamination due to failed library preparation

*DOI: http://dx.doi.org/10.5772/intechopen.105929*

## 5. Quantification—Feature counts

Once the alignment QC is done and looks satisfactory, the next step is to use the annotation of the reference organism to quantify which genes have a heightened expression or reduced expression. This is the step that is going to help perform the following analyses. This is also the part where the effect of strandedness can be observed.

## 6. Sample correlation and principal components

Once the counts are obtained, the subsequent steps are to interrogate whether there are any clusters observed among the various samples.

Ideally, the correlation and principal components (as a by-product) should tell you if the replicates sequenced are clustering together. Are there any batch effects or other covariates to be adjusted for?

A simple Pearson correlation, which explores a linear relationship, would tell on a scale of -1 (no)—0 (bad)—1 (good) correlation in the sample.

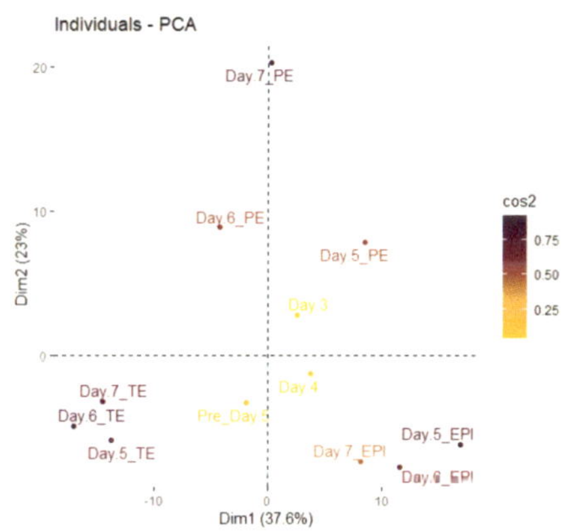


PCA plots can tell which individuals here have similar sets of gene expression profile

## 7. Comparative analyses—DESeq2: Volcano plots, heatmaps, etc.

Depending on the experiments planned, you prepare metadata stating what the datasets represent. Are the samples differing based on treatments, populations, time points, etc.? Once the objective is understood, you tend to model the formula (called design) in DESeq2, which looks something like:

design= ~ condition + age

design= ~ batch + condition

The thing to note here is that the variable immediately after "~" is called the controlling feature, and after the "+" is the effecting feature. You want to control things such as sequencing batches, populations, genders, etc., while exploring the impact of the experiment on the condition/age, etc. In the absence of the + sign, the first feature gets explored.

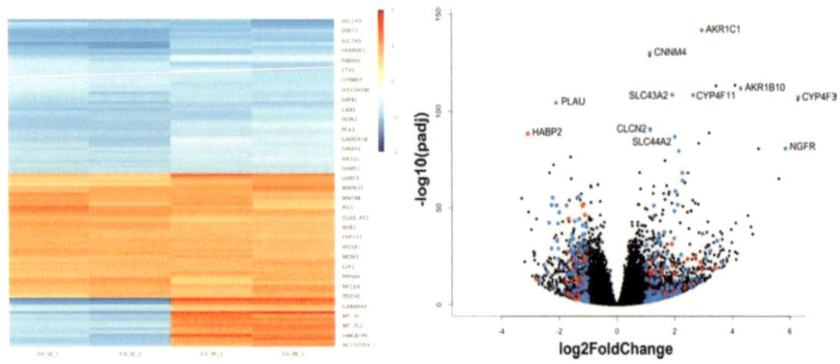


Heatmap of the top "N" differential genes and volcano plot of the gene expression

Genes in red are positively regulated while in blue are negative. Genes can also be clustered hierarchically based on expression patterns to see which genes are expressing together

A volcano plot tells about the expression modulation in the context of the confidence intervals. A general threshold that people use for the confidence interval is 0.05 (adjusted p-values) or false discovery rates. This is drawn on the outcome reported by DESeq2 using the log of fold changes vs negative log of the P-adjusted values/false discovery rates.

DESeq2 is one of the approaches, which uses regularized logarithm transformation to normalize the counts. People also use variance stabilizing transform, TMM, UQ, etc.

## 8. Term enrichments—GO overrepresentation, reactome and KEGG pathways, etc

One of the end goals of an RNA-Seq study is to generally understand biologically what the perturbations are. These can be cell cycle disruption, increased metabolic processes, cell senescence, cell growth, increased trafficking of vesicles, etc. A few tools to perform these are: Profiler, cluster profile, etc.

Reactome, KEGG, etc., are databases that are created by understanding physiologically how the genes are arranged and flow in a pathway, or transfer them from systems for which we have already understood them.

GO/gene ontology is similarly a technique that is universally applied to the tree of life. A GO term for a gene is assigned after studying the function of the gene and then assigned a generalized role to it.

By figuring out genes that have a similar role in the expressions, and their modulations, we can try to understand what would have been the impact of these genes on the organism.

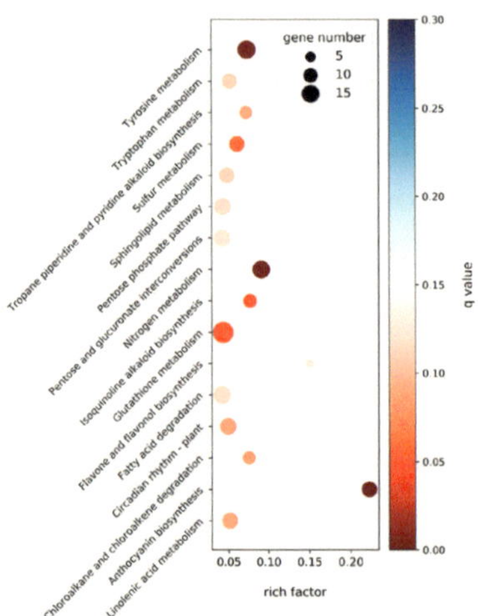

KEGG pathways. The colored circles are scaled according to the number of genes involved in the particular pathway. The q value tells the confidence of the assignment. Rich factor is a ratio of genes observed in the study/genes seen in the pathway.

In this case, ~20 genes would be involved in glutathione metabolism with a q value of 0, showing that this pathway is being impacted.

While KEGG/reactome/GOs are shown as an example, one can create their versions of databases and try to compute the impact of the genes observed accordingly.

## Author details

Nakul D. Magar[1†], Priya Shah[2†], K. Harish[3], Tejas C. Bosamia[4], Kalyani M. Barbadikar[1*], Yogesh M. Shukla[5], Amol Phule[6], Harshvardhan N. Zala[7], Maganti Sheshu Madhav[1], Satendra Kumar Mangrauthia[1], Chirravuri Naga Neeraja[1] and Raman Meenakshi Sundaram[1]

1 Biotechnology Section, ICAR-Indian Institute of Rice Research, Hyderabad, India

2 International Crops Research Institute for the Semi-Arid Tropics, Hyderabad, India

3 Ikigai Informatics, Vadodara, India

4 Plant Omics Division, CSIR-Central Salt and Marine Chemicals Research Institute, Bhavnagar, India

5 Faculty of Agriculture, Anand Agricultural University, Anand, India

6 ICAR-Indian Institute of Rice Research, Hyderabad, India

7 College of Basic Science and Humanities, Sardarkrushinagar Dantiwada Agricultural University, Sardarkrushinagar, India

*Address all correspondence to: kalyaniaau@gmail.com; kalyani.mb@icar.gov.in

† Equal contribution.

## References

[1] Schena M, Shalon D, Davis RW, Brown PO. Quantitative monitoring of gene expression patterns with a complementary DNA microarray. Science. 1995;**270**:467-470

[2] Velculescu VE, Zhang L, Vogelstein B, Kinzler KW. Serial analysis of gene expression. Science. 1995;**270**:484-487

[3] Chu Y, Corey DR. RNA sequencing: Platform selection, experimental design, and data interpretation. Nucleic Acid Therapeutics. 2012;**22**:271-274

[4] Wilhelm BT, Marguerat S, Watt S, Schubert F, Wood V, Goodhead I, et al. Dynamic repertoire of a eukaryotic transcriptome surveyed at single-nucleotide resolution. Nature. 2008;**453**:1239-1243

[5] Metzker ML. Sequencing technologies—the next generation. Nature Reviews: Genetics. 2010;**11**:31-46

[6] Wang Z, Gerstein M, Snyder M. RNA-Seq: A revolutionary tool for transcriptomics. Nature Reviews: Genetics. 2009;**10**:57-63

[7] Ozsolak F, Milos PM. RNA sequencing: Advances, challenges and opportunities. Nature Reviews: Genetics. 2011;**12**:87-98

[8] Marguerat S, Bähler J. RNA-seq: From technology to biology. Cellular and Molecular Life Sciences. 2010;**67**:569-579

[9] Martin JA, Wang Z. Next-generation transcriptome assembly. Nature Reviews. Genetics. 2011;**12**:671-682

[10] Shendure J, Ji H. Next-generation DNA sequencing. Nature Biotechnology. 2008;**26**:1135-1145

[11] Guo Y, Li J, Li C-I, Long J, Samuels DC, Shyr Y. The effect of strand bias in Illumina short-read sequencing data. BMC Genomics. 2012;**13**:1-11

[12] Bahassi EM, Stambrook PJ. Next-generation sequencing technologies: Breaking the sound barrier of human genetics. Mutagenesis. 2014;**29**:303-310

[13] Mikheyev AS, Tin MMY. A first look at the Oxford Nanopore MinION sequencer. Molecular Ecology Resources. 2014;**14**:1097-1102

[14] Maitra RD, Kim J, Dunbar WB. Recent advances in nanopore sequencing. Electrophoresis. 2012;**33**:3418-3428

[15] Lee JH, Daugharthy ER, Scheiman J, Kalhor R, Yang JL, Ferrante TC, et al. Highly multiplexed subcellular RNA sequencing in situ. Science. 2014;**343**:1360-1363

[16] Ayub M, Bayley H. Individual RNA base recognition in immobilized oligonucleotides using a protein nanopore. Nano Letters. 2012;**12**:5637-5643

[17] Milward EA, Daneshi N, Johnstone DM. Emerging real-time technologies in molecular medicine and the evolution of integrated 'pharmacomics' approaches to personalized medicine and drug discovery. Pharmacology & Therapeutics. 2012;**136**(3):295-304. DOI: 10.1016/j.pharmthera.2012.08.008

[18] Piepenburg O, Williams CH, Stemple DL, Armes NA. DNA detection using recombination proteins. PLoS Biology. 2006;**4**:e204

[19] Parkinson J, Blaxter M. Expressed sequence tags: An overview. Expressed Sequence Tags. 2009:1-12

[20] Hatey F, Tosser-Klopp G, Clouscard-Martinato C, Mulsant P, Gasser F. Expressed sequence tags for genes: A review. Genetics, Selection, Evolution. 1998;**30**:521-541

[21] Lopez C, Soto M, Restrepo S, Piégu B, Cooke R, Delseny M, et al. Gene expression profile in response to Xanthomonas axonopodis pv. manihotis infection in cassava using a cDNA microarray. Plant Molecular Biology. 2005;**57**:393-410

[22] Yonekura-Sakakibara K, Saito K. Functional genomics for plant natural product biosynthesis. Natural Product Reports. 2009;**26**:1466-1487

[23] Kalia RK, Rai MK, Kalia S, Singh R, Dhawan AK. Microsatellite markers: An overview of the recent progress in plants. Euphytica. 2011;**177**:309-334

[24] Paterson AH, Bowers JE, Burow MD, Draye X, Elsik CG, Jiang C-X, et al. Comparative genomics of plant chromosomes. Plant Cell. 2000;**12**:1523-1539

[25] Ewing RM, Claverie JM. EST databases as multi-conditional gene expression datasets. Biocomput. World Scientific. 1999;**200**:430-432

[26] Ogihara Y, Mochida K, Nemoto Y, Murai K, Yamazaki Y, Shin-I T, et al. Correlated clustering and virtual display of gene expression patterns in the wheat life cycle by large-scale statistical analyses of expressed sequence tags. The Plant Journal. 2003;**33**:1001-1011

[27] Morozova O, Hirst M, Marra MA. Applications of new sequencing technologies for transcriptome analysis. Annual Review of Genomics and Human Genetics. 2009;**10**:135-151

[28] Diatchenko L, Lau YF, Campbell AP, et al. Suppression subtractive hybridization: A method for generating differentially regulated or tissue-specific cDNA probes and libraries. Proceedings of the National Academy of Sciences. 1996;**93**(12):6025-6030

[29] Shiraki T, Kondo S, Katayama S, Waki K, Kasukawa T, Kawaji H, et al. Cap analysis gene expression for high-throughput analysis of transcriptional starting point and identification of promoter usage. Proceedings of the National Academy of Sciences. 2003;**100**:15776-15781

[30] Southern E, Mir K, Shchepinov M. Molecular interactions on microarrays. Nature Genetics. 1999;**21**:5-9

[31] Daudén E, Farmacogenética II. Métodos moleculares de estudio, bioinformática y aspectos éticos. Actas Dermo-Sifiliográficas. 2007;**98**:3-13

[32] Malone JH, Oliver B. Microarrays, deep sequencing and the true measure of the transcriptome. BMC Biology. 2011;**9**:1-9

[33] Santos CA, Blanck DV, de Freitas PD. RNA-seq as a powerful tool for penaeid shrimp genetic progress. Frontiers in Genetics. 2014;**5**:298

[34] San Segundo-Val I, Sanz-Lozano CS. Introduction to the gene expression analysis. Molecular Genetics in Asthma. 2016;**2016**:29-43

[35] Lowe R, Shirley N, Bleackley M, Dolan S, Shafee T. Transcriptomics technologies. PLoS Computational Biology. 2017;**13**:e1005457

[36] de Vienne D. Molecular Markers in Plant Genetics and Biotechnology. CRC Press; 2003

[37] Pandey V, Nutter RC, Prediger E. Applied biosystems SOLiD™ System:

Ligation-Based Sequencing. In: Next Generation Genome Sequencing: Towards Personalized Medicine. Wiley. 2008:29-41.

[38] Edwards M. Whole-genome Sequencing for Marker Discovery. In: Henry RJ, editor. Molecular Markers in Plants. Oxford: Blackwell Publishing Ltd.; 2012:21-34

[39] Schadt EE, Turner S, Kasarskis A. A window into third-generation sequencing. Human Molecular Genetics. 2010;**19**:R227-R240

[40] Margulies M, Egholm M, Altman WE, Attiya S, Bader JS, Bemben LA, et al. Genome sequencing in microfabricated high-density picolitre reactors. Nature. 2005;**437**:376-380

[41] McKernan KJ, Peckham HE, Costa GL, McLaughlin SF, Fu Y, Tsung EF, et al. Sequence and structural variation in a human genome uncovered by short-read, massively parallel ligation sequencing using two-base encoding. Genome Research. 2009;**19**:1527-1541

[42] Amarasinghe SL, Su S, Dong X, Zappia L, Ritchie ME, Gouil Q. Opportunities and challenges in long-read sequencing data analysis. Genome Biology. 2020;**21**:1-16

[43] Levin JZ, Yassour M, Adiconis X, Nusbaum C, Thompson DA, Friedman N, et al. Comprehensive comparative analysis of strand-specific RNA sequencing methods. Nature Methods. 2010;**7**:709-715

[44] Efroni I, Ip P-L, Nawy T, Mello A, Birnbaum KD. Quantification of cell identity from single-cell gene expression profiles. Genome Biology. 2015;**16**:1-12

[45] Trapnell C, Hendrickson DG, Sauvageau M, Goff L, Rinn JL, Pachter L. Differential analysis of gene regulation at transcript resolution with RNA-seq. Nature Biotechnology. 2013;**31**:46-53

[46] DeLuca DS, Levin JZ, Sivachenko A, Fennell T, Nazaire M-D, Williams C, et al. RNA-SeQC: RNA-seq metrics for quality control and process optimization. Bioinformatics. 2012;**28**:1530-1532

[47] Wang L, Wang S, Li W. RSeQC: Quality control of RNA-seq experiments. Bioinformatics. 2012;**28**:2184-2185

[48] Ryan MC, Cleland J, Kim R, Wong WC, Weinstein JN. SpliceSeq: A resource for analysis and visualization of RNA-Seq data on alternative splicing and its functional impacts. Bioinformatics. 2012;**28**:2385-2387

[49] Xu G, Deng N, Zhao Z, Judeh T, Flemington E, Zhu D. SAMMate: A GUI tool for processing short read alignments in SAM/BAM format. Source Code for Biology and Medicine. 2011;**6**:1-11

[50] Tang S, Riva A. PASTA: Splice junction identification from RNA-Sequencing data. BMC Bioinformatics. 2013;**14**:1-11

[51] Chen LY, Wei K-C, Huang AC-Y, Wang K, Huang C-Y, Yi D, et al. RNASEQR—a streamlined and accurate RNA-seq sequence analysis program. Nucleic Acids Research. 2012;**40**:e42-e42

[52] Kim D, Pertea G, Trapnell C, Pimentel H, Kelley R, Salzberg SL. TopHat2: Accurate alignment of transcriptomes in the presence of insertions, deletions and gene fusions. Genome Biology. 2013;**14**:1-13

[53] Wang K, Singh D, Zeng Z, Coleman SJ, Huang Y, Savich GL, et al. MapSplice: Accurate mapping of RNA-seq reads for splice junction discovery. Nucleic Acids Research. 2010;**38**:e178-e178

[54] Dimon MT, Sorber K, DeRisi JL. HMMSplicer: A tool for efficient and sensitive discovery of known and novel splice junctions in RNA-Seq data. PLoS One. 2010;**5**:e13875

[55] Wu TD, Reeder J, Lawrence M, Becker G, Brauer MJ. GMAP and GSNAP for Genomic Sequence Alignment: Enhancements to Speed, Accuracy, and Functionality. Springer; 2016. pp. 283-334

[56] Bao H, Xiong Y, Guo H, Zhou R, Lu X, Yang Z, et al. MapNext: A software tool for spliced and unspliced alignments and SNP detection of short sequence reads. BMC Genomics. 2009;**10**:1-6

[57] Dobin A, Davis CA, Schlesinger F, Drenkow J, Zaleski C, Jha S, et al. STAR: Ultrafast universal RNA-seq aligner. Bioinformatics. 2013;**29**:15-21

[58] Schulz MH, Zerbino DR, Vingron M, Birney E. Oases: Robust de novo RNA-seq assembly across the dynamic range of expression levels. Bioinformatics. 2012;**28**:1086-1092

[59] Xie Y, Wu G, Tang J, Luo R, Patterson J, Liu S, et al. SOAPdenovo-Trans: de novo transcriptome assembly with short RNA-Seq reads. Bioinformatics. 2014;**30**:1660-1666

[60] Grabherr MG, Haas BJ, Yassour M, Levin JZ, Thompson DA, Amit I, et al. Trinity: Reconstructing a full-length transcriptome without a genome from RNA-Seq data. Nature Biotechnology. 2011;**29**:644

[61] Robertson G, Schein J, Chiu R, Corbett R, Field M, Jackman SD, et al. De novo assembly and analysis of RNA-seq data. Nature Methods. 2010;**7**:909-912

[62] Hardcastle TJ, Kelly KA. baySeq: Empirical Bayesian methods for identifying differential expression in sequence count data. BMC Bioinformatics. 2010;**11**:1-14

[63] Wang L, Feng Z, Wang X, Wang X, Zhang X. DEGseq: An R package for identifying differentially expressed genes from RNA-seq data. Bioinformatics. 2010;**26**:136-138

[64] Anders S, Huber W. Differential expression of RNA-Seq data at the gene level–the DESeq package. European Molecular Biology Lab. 2012;**10**:f1000

[65] Robinson MD, McCarthy DJ, Smyth GK. edgeR: A Bioconductor package for differential expression analysis of digital gene expression data. Bioinformatics. 2010;**26**:139-140

[66] Beier S, Thiel T, Münch T, Scholz U, Mascher M. MISA-web: A web server for microsatellite prediction. Bioinformatics. 2017;**33**:2583-2585

[67] Griffith M, Griffith OL, Mwenifumbo J, Goya R, Morrissy AS, Morin RD, et al. Alternative expression analysis by RNA sequencing. Nature Methods. 2010;**7**:843-847

[68] Singh D, Orellana CF, Hu Y, Jones CD, Liu Y, Chiang DY, et al. FDM: A graph-based statistical method to detect differential transcription using RNA-seq data. Bioinformatics. 2011;**27**:2633-2640

[69] Drewe P, Stegle O, Hartmann L, Kahles A, Bohnert R, Wachter A, et al. Accurate detection of differential RNA processing. Nucleic Acids Research. 2013;**41**:5189-5198

[70] Shi Y, Jiang H. rSeqDiff: Detecting differential isoform expression from RNA-Seq data using hierarchical likelihood ratio test. PLoS One. 2013;**8**:e79448

[71] Li B, Dewey CN. RSEM: Accurate transcript quantification from RNA-Seq data with or without a reference genome. BMC Bioinformatics. 2011;**12**:1-16

[72] Nicolae M, Mangul S, Măndoiu II, Zelikovsky A. Estimation of alternative splicing isoform frequencies from RNA-Seq data. Algorithms for Molecular Biology. 2011;**6**:1-13

[73] Glaus P, Honkela A, Rattray M. Identifying differentially expressed transcripts from RNA-seq data with biological variation. Bioinformatics. 2012;**28**:1721-1728

[74] Martin J, Bruno VM, Fang Z, Meng X, Blow M, Zhang T, et al. Rnnotator: An automated de novo transcriptome assembly pipeline from stranded RNA-Seq reads. BMC Genomics. 2010;**11**:1-8

[75] Sacomoto GAT, Kielbassa J, Chikhi R, Uricaru R, Antoniou P, Sagot M-F, et al. K is s plice: de-novo calling alternative splicing events from rna-seq data. BMC Bioinformatics. 2012;**13**:1-12

[76] Sherman BT, Tan Q, Collins JR, Alvord WG, Roayaei J, Stephens R, et al. The DAVID Gene Functional Classification Tool: A novel biological module-centric algorithm to functionally analyze large gene lists. Genome Biology. 2007;**8**:1-16

[77] Huang DW, Sherman BT, Zheng X, Yang J, Imamichi T, Stephens R, et al. Extracting biological meaning from large gene lists with DAVID. Current Protocol Bioinforma. 2009;**27**:1-13

[78] Subramanian A, Kuehn H, Gould J, Tamayo P, Mesirov JP. GSEA-P: A desktop application for Gene Set Enrichment Analysis. Bioinformatics. 2007;**23**:3251-3253

[79] El-Metwally S, Ouda OM, Helmy M. New horizons in next-generation sequencing. Next Generation. 2014;**2014**:51-59

[80] Shen W, Li H, Teng R, Wang Y, Wang W, Zhuang J. Genomic and transcriptomic analyses of HD-Zip family transcription factors and their responses to abiotic stress in tea plant (Camellia sinensis). Genomics. 2019;**111**:1142-1151

[81] Leisner CP, Yendrek CR, Ainsworth EA. Physiological and transcriptomic responses in the seed coat of field-grown soybean (Glycine max L. Merr.) to abiotic stress. BMC Plant Biology. 2017;**17**:1-11

[82] Kreszies T, Shellakkutti N, Osthoff A, Yu P, Baldauf JA, Zeisler-Diehl VV, et al. Osmotic stress enhances suberization of apoplastic barriers in barley seminal roots: Analysis of chemical, transcriptomic and physiological responses. The New Phytologist. 2019;**221**:180-194

[83] Barrett T, Troup DB, Wilhite SE, Ledoux P, Rudnev D, Evangelista C, et al. NCBI GEO: Mining tens of millions of expression profiles—database and tools update. Nucleic Acids Research. 2007;**35**:D760-D765

[84] Burch-Smith TM, Anderson JC, MartinGB, Dinesh-KumarSP. Applications and advantages of virus-induced gene silencing for gene function studies in plants. The Plant Journal. 2004;**39**(5):734-746

[85] Fang Z, Cui X. Design and validation issues in RNA-seq experiments. Briefings in Bioinformatics. 2011;**12**:280-287

[86] Phule AS, Barbadikar KM, Madhav MS, Senguttuvel P, Babu MBB, Ananda KP. Genes encoding membrane proteins showed stable expression in rice under aerobic condition: Novel set of

reference genes for expression studies. 3 Biotech. 2018;**8**:1-12

[87] Bustin SA, Benes V, Garson JA, Hellemans J, Huggett J, Kubista M, et al. The MIQE Guidelines: Minimum I nformation for Publication of Quantitative Real-Time PCR Experiments, Clinical Chemistry. 1 April 2009;**55**(4):611-622. DOI: 10.1373/clinchem.2008.112797

[88] Schmittgen TD, Livak KJ. Analyzing real-time PCR data by the comparative CT method. Nature Protocols. 2008;**3**:1101-1108

[89] Sripathi VR, Anche VC, Gossett ZB, Walker LT. Recent applications of RNA sequencing in food and agriculture. In: Louis IV, editor. Applications of RNA-Seq in Biology and Medicine. London: IntechOpen; 2021

[90] Byron A, Van Keuren-Jensen K, Engelthaler D. Translating RNA sequencing into clinical diagnostics: Opportunities and challenges. Nature Review Genetics. 2016;**17**:257-271

[91] Li X, Wang CY. From bulk, single-cell to spatial RNA sequencing. International Journal of Oral Science. 2021;**13**:36

[92] Zhao L, Zhang H, Kohnen MV, Prasad KVSK, Gu L, Reddy ASN. Analysis of transcriptome and epitranscriptome in plants using PacBio Iso-Seq and nanopore-based direct RNA sequencing. Frontiers in Genetics. 2019;**10**:253

[93] Westermann AJ, Barquist L, Vogel J. Resolving host–pathogen interactions by dual RNA-seq. PLoS Pathogens. 2017;**13**(2):e1006033

[94] Wainberg M, Sinnott-Armstrong N, Mancuso N, et al. Opportunities and challenges for transcriptome-wide association studies. Nature Genetics. 2019;**51**:592-599

[95] Gardini A. Global Run-On Sequencing (GRO-Seq). Methods in Molecular Biology. 2017;**1468**:111-120

Section 2

# Gene Expression and Human Diseases

Chapter 6

# *APC* and *MSH2* mRNA Quantitative Gene Expression and Bayesian Analysis of Proband in Hereditary Colorectal Carcinoma

*Tjahjadi Robert Tedjasaputra, Mochammad Hatta, Muhammad Nasrum Massi, Rosdiana Natzir, Ilhamjaya Patellongi, Marcellus Simadibrata, Rina Masadah, Muhammad Luthfi Parewangi, Prihantono, Andi Asadul Islam, Agussalim Bukhari, Rinda Nariswati, Shirly Elisa Tedjasaputra, Vincent Tedjasaputra and Jonathan Salim*

## Abstract

Heterozygote relatives have approximately 80% lifetime colorectal cancer (CRC) risk. mRNA gene expression and Bayesian theorem can calculate CRC's family risk through the initial pedigree proportion appended with conditional information. The study is the first to report such an application. The present cross-sectional and translational investigation tracked CRC patients' tissue and blood measurement of adenomatous polyposis coli (*APC*) and MutS homolog (*MSH*)*2 mRNA* quantitative gene expressions, control matching, and ancestral analysis by pedigree and Bayesian theorem. Among 40 CRC patients, mean tissue level and hereditary cutoff of APC are 13,261 (670) fold-change (fc) and 12,195 fc, while 12,219 (756) fc and 11,059 fc for MSH2. A quarter of the CRC patients had a history of familial CRC. Meanwhile, four CRC patients and 10 probands were evaluated for recurrence risk via pedigree, quantitative PCR, and Bayesian analysis. We determined a cutoff point for hereditary mRNA quantitative expression. *APC* and *MSH2* levels in the CRC subjects were significantly lower than controls. The Bayesian analysis builds ways to calculate relative risk in CRC patients' family members and application in clinical practice.

**Keywords:** hereditary CRC, APC gene, MSH gene, Bayesian analysis

## 1. Introduction

The continuing morbidity and lethality of colorectal cancer (CRC) do not always stop at the diseased person. In fact CRC holds the top place in familial inherited case prevalence [1]. Hereditary CRC with clear-cut forms overall can be divided into lynch syndrome (LS) or hereditary non-polyposis colorectal cancer (HNPCC) and familial adenomatous polyposis (FAP), which are inherited by autosomal dominant pattern [2]. Those who related to parents or grandparents with an autosomal dominant trait has at least an 80% chance for lifetime risk of CRC incidence [3].

Screening for cancers is dependent on every individual. Those without any familial cancer history can start colonoscopy for CRC screening at 50 years old. Nevertheless, the age is smaller by a magnitude if you have a CRC first-degree relative. A CRC individual will raise your risk by two to three times more than normal; however, more relatives with the disease may equal to an exponential risk increase [4].

Familial characteristics such as age, disease onset, size, and health history often pose precarious conditions to the internist and gastroenterologist who did CRC hereditary screening by the Amsterdam and Bethesda criteria. These were illustrated in the low guidelines' performances from both Revised Bethesda Guideline and Amsterdam II Criteria against molecular tumor analysis with 50 and 25% sensitivity and 7 and 38% positive predictive values [5]. We accordingly need a more swift and stable method to screen for hereditary CRC, such as with the implementation of family history, molecular expression, and Mendel inheritance concept [6, 7].

The Mendel hereditary concept is well-performed in screening or determining autosomal and gonadal patterns risks, as it can compute recurrence probability; however, it cannot be quickly adjusted for mutation, external factors, and coverage changes, since it focused more on empirical recurrences. Yet the application of only such analysis is questionable, as most traits are not generalizable. Hence, family members' recurrence risk should be calculated with prior Mendelian risk and geared with personal genotyping and environmental conditional probability [8–11].

Genetic studies are becoming more present recently with research around genetic matters like DNA sequencing or polymorphism [12]. They open up a new horizon for disease susceptibility and inheritance analysis, including malignancy. However, RNA study was still rare as mistakes in the nucleotide base or elsewhere will be quickly dealt by the proofreading and the mismatch repair (MMR) genes [13].

Adenomatous polyposis coli (*APC*) gene exhibited a unique causal relationship to the incidence of hereditary FAP from mutation on the fifth chromosome's second region and first band. LS conversely rises from mutations in the second, third, and seventh chromosome of several different genes, including human MutL homolog 1, human MutS homolog 2 and 6, as well as human post-meiotic segregation 1 and 2 (*hMLH1*, *hMSH2*, *hMSH6*, *hPMS1*, and *hPMS2*) [14].

In commencing the current study, research operators or the authors need to be more aware of their surroundings. This time, huge complex calculations and uncommon Bayesian prior and posterior analysis were implemented. There is no former report on the *APC* and *MSH2* genomic RNA expressions to CRC risk with modified Bayesian estimation per the authors' knowledge [15–17]. We hoped the current study was able to officialize an adequate hereditary measure through gene expression and the families able to incorporate Bayesian into their risk of CRC.

## 2. Materials and methods

### 2.1 Study design

The current translational study adopted a cross-sectional design in assessing 71 subjects from May 2018 to December 2019. Medical Ethics Committee of Hasanuddin University ensured the research commencement had followed Helsinki declaration and institutional review board (IRB) standards with certification of 884/H4.8.45.31/PP31-Komite/2018. Every subject had understood and agreed to participate as shown by the signed informed consent form. The author priorly measured the minimum sample size by 5% alpha and 80% power.

### 2.2 Subject enrollment

Subjects were consecutively gathered from Tarakan General Hospital in Jakarta and Siloam Hospitals Lippo Village in Tangerang. The case group broadly enrolled all 41 CRC patients who had undergone a biopsy in either hospital. Gastroenterologists and oncologists made the CRC diagnosis based on the clinical symptoms, physical examinations, and supporting investigations (i.e., endoscopy and histopathological findings). We contrarily enlisted normal patients or CRC patients' relatives who had been matched by age, sex, and body mass index to the control group.

Exclusion of patients from either group may happen at any time of the study if they had: (1) presence or history of other malignancies or inflammatory bowel disease, (2) ever done chemotherapy or radiotherapy, (3) illnesses that inhibit communication, and (4) refuse to participate.

### 2.3 Data and sample collection

The current study investigated *APC* and *MSH2* quantitative genotypic expressions as well as hereditary possibilities. 0.3 ml of blood samples were taken from all 71 subjects using one cc syringes, yet only CRC subjects did colonoscopy biopsy. Each of the blood and tissue samples was then laced into separated sample tubes containing L6 buffer preservative, specifically created by Hasanuddin University from a slightly modified version of the buffer in the Boom RNA extraction method. **Figure 1** showed the complete RNA extraction techniques by the NucleoSpin technique (#740200.50) [18]. The isolated extraction results were subsequently amplified using a real-time PCR

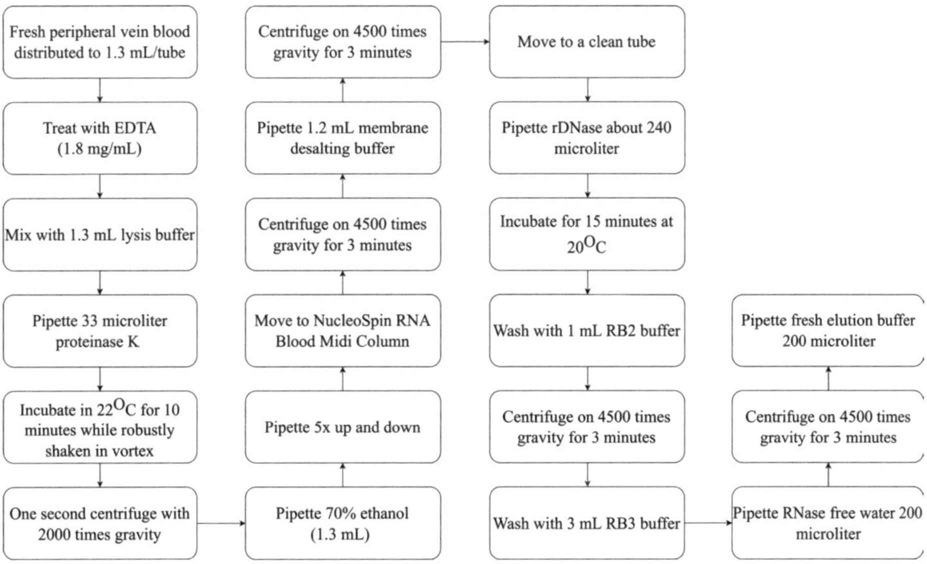


**Figure 1.**
*RNA extraction technique. Step by step pathway of the RNA extraction with NucleoSpin technique (#740200.50) [18].*

| Gene | Orientation | Sequence | Primer [19, 22] |
|---|---|---|---|
| Specific primer target[a] | | | |
| *MSH2* | Forward | 5′ to 3′ | CATCCAGGCATGCTTGTGTTGA |
| | Reverse | | GCAGTCCACAATGGACACTTC |
| *APC* | Forward | | TGTCCCTCCGTTCTTATGGAA |
| | Reverse | | TCTTGGAAATGAACCCATAGGAA |
| Internal control genes[a] | | | |
| ß-actin | Forward | 5′ to 3′ | ACAGAGCCTCGCCTTTGCCGAT |
| | Reverse | | CTTGCACATGCCGGAGCCGTT |
| *GAPDH* | Forward | | CGCTCTCTGCTCCTCCTGTT |
| | Reverse | | CCATGGTGTCTGAGCGATGT |

[a]*Designed by Macrogen (Seoul, Korea).*

**Table 1.**
*Genetic primers.*

(RT-PCR) and then measured with a Bio-Rad CFX Manager 3.1 (Bio-Rad Laboratories, Hercules, USA) [19–21]. After that, we also applied the Bayesian probability analysis on the probands' age, *APC*, and *MSH2* data to yield CRC risk estimations.

## 2.4 Polymerase chain reaction (PCR)

The present study used RT-PCR to detect the mRNA expression of *MSH2* and *APC* genes with the following primers (**Table 1**). First, they entered the initial denaturing phase with 94°C for 3 minutes. Then the process continued with 38 cycles of annealing stage in 54°C for 30 s and extension stage in 72°C for 30–40 s [19–21]. Note that each gene has its unique amplicon length. For example, *APC* is 89 bp long, 81 bp for *GAPDH*, 215 bp for *MSH2*, and 109 bp for ß-actin [23].

We procured the RT-PCR materials from Power SYBR Green PCR Master Mix (Applied Biosystems, Foster City, USA). Additionally, we used CFX Connect real-time PCR system from Bio-Rad Laboratories for the measurement [22, 23].

## 2.5 Bayesian probability

The current investigation quantified the CRC risk among family members through Bayesian analysis of the Mendelian hereditary, genetic, and direct mutations data [4, 16]. We derived a posterior probability equation (Eq. (1)) to estimate the family CRC risk from the coupling of the conditional and prior probability theorems.

$$p(A|B) = \frac{p(B|A)p(A)}{p(B)} \text{ and } E(\beta) = \int \beta \, p(\beta|\gamma|,\vartheta)d\beta$$

$$p(\beta|\gamma|,n) = \frac{p(\gamma,\beta,|\vartheta)}{p(\gamma|\vartheta)} = \frac{p(\gamma,\beta|\vartheta)}{\int p(\gamma,u|\vartheta)du} = \frac{f(\gamma|\beta)\varpi(\beta|\vartheta)}{\int f(\gamma|u)\varpi(u|\vartheta)du} \quad (1)$$

### 2.6 Statistical analysis

Statisticians used descriptive statistics to univariately portray the subjects' demographic characteristics and histological findings, as opposed to Shapiro-Wilk for determining the numeric data normality. Further bivariate analysis of the parametric numeric variables used *t*-test, while non-parametric used Mann-Whitney. Meanwhile, they tested categorical variables by either $x^2$ or Fisher exact. Bayesian analysis was employed last for adjusted estimation of the CRC risk in the probands (i.e., relatives of the hereditary CRC patients).

## 3. Results

Forty CRC patients and 31 healthy controls had a 100% participation rate within the research period. Those with CRC on average live 5.19 years longer with 1.21 kg/$m^2$ lower body mass index (BMI) than the counterpart. The sex difference was also apparent with a 1.11:1 vs. 0.72:1 male to female ratio among the case and control groups, respectively. We also determined the cellular differentiation levels among the biopsied CRC subjects, with results in **Table 2**.

There was a significant difference in blood *APC* levels between CRC and control subjects. There was a lower mean value of *MSH2* in CRC but no substantial difference between CRC and control subjects because of the outlier (**Table 3**).

Hereditary screening of the CRC subjects came next. Analyzing the CRC risk with Bayesian Analysis is futile if the disease is not hereditary in the first place. There are however no prior validated data on cutoff amounts for CRC hereditary trait from *APC* and *MSH2* gene expression. Hence, we established the required cutoff values through the fifth percentile technique.

| Variable | Subject group | | p |
|---|---|---|---|
| | CRC ($n$ = 40) | Control ($n$ = 31) | |
| Age (year)[a] | 56.80 (8.40) | 51.61 (13.44) | |
| Body mass index (kg/$m^2$)[a] | 22.41 (3.29) | 23.62 (3.41) | |
| Sex[b] | | | >0.05 |
| • Male | 21 (52.5) | 13 (41.9) | |
| • Female | 19 (47.5) | 18 (58.1) | |
| Cellular differentiation[b] | | — | — |
| • Adenocarcinoma | | | |
| • Well | 26 (65.0) | | — |
| Fair | 6 (15.0) | | |
| Poor | 7 (17.5) | | |
| Neuroendocrine carcinoma | 1 (2.5) | | |

[a]*Mean (standard deviation).*
[b]*n (%).*

**Table 2.**
*Baseline characteristics.*

| Gene | Subject group (fold-change) | | p |
|---|---|---|---|
| | CRC ($n$ = 40) | Control ($n$ = 31) | |
| Blood sample | | | |
| *APC* | | | |
| • Median (range) | 12,156.5 (5848–15,035) | 13,260 (12,080–14,376) | 0.014 |
| • Mean (SD) | 11,578.68 (2638.23) | 13,261.74 (670.56) | 0.014 |
| *MSH2* | | | |
| • Median (range) | 12,554.5 (4230–14,559) | 12,146 (11,029–13,633) | 0.116 |
| • Mean (SD) | 11,411.05 (2912.45) | 12,219.87 (756.87) | 0.465 |
| Tissue sample | | | |
| *APC* | | — | — |
| • Median (range) | 8337.0 (5060–13,087) | | |
| • Mean (SD) | 8147.78 (1875.12) | | |
| *MSH2* | | — | — |
| • Median (range) | 7485.0 (4174–14,218) | | |
| • Mean (SD) | 7475.20 (1946.24) | | |

**Table 3.**
*APC and* MSH2 *gene expression between groups.*

**Table 4** showed the percentiles distribution of both *APC* and *MSH2* quantitative expressions among the control group. The fifth percentile of both genes adequately fits to be hereditary cutoff values since it had no significant difference to the first and third percentile. Henceforth, hereditary CRC was very likely in those with over 12,195.80 fc *APC* or 11,059.60 fc *MSH2*.

The CRC subjects were then distributed nicely into either the hereditary or sporadic category with the determined cutoff. Gene expressions equal to and above the cutoff positively correspond to a hereditary status. A majority proportion (52.5%) of the 40 people with colorectal cancer had hereditary nature based on both *APC* and *MSH2* cutoffs. Nonetheless, the hereditary rate decreased by 2.5% and 20.0% if only using cutoff from either one (**Table 5**).

Complete pedigree analysis of the family age, health, gender, and family history of diseases is essential for the estimation of CRC risk. There were merely eight subjects with positive CRC in the family and even then, half were dropped because of vague recollection or retracted permission. We consequently extracted only 10 probands from the four CRC families for Bayesian analysis (**Figure 2A–D**).

| Percentile | Gene expression ($n$ = 31) | |
|---|---|---|
| | *APC* (fc) | *MSH2* (fc) |
| First | 12,080.00 | 11,029.00 |
| Third | 12,080.00 | 11,029.00 |
| Fifth | 12,195.80 | 11,059.60 |

**Table 4.**
*Gene expression percentile distribution.*

| Cutoff gene | Hereditary (*n* (%)) | Sporadic (*n* (%)) |
|---|---|---|
| *APC* | 20 (50.0) | 20 (50.0) |
| *MSH2* | 13 (32.5) | 27 (67.5) |
| *APC* and *MSH2* | 21 (52.5) | 19 (47.5) |

**Table 5.**
*CRC subjects' hereditary distribution.*

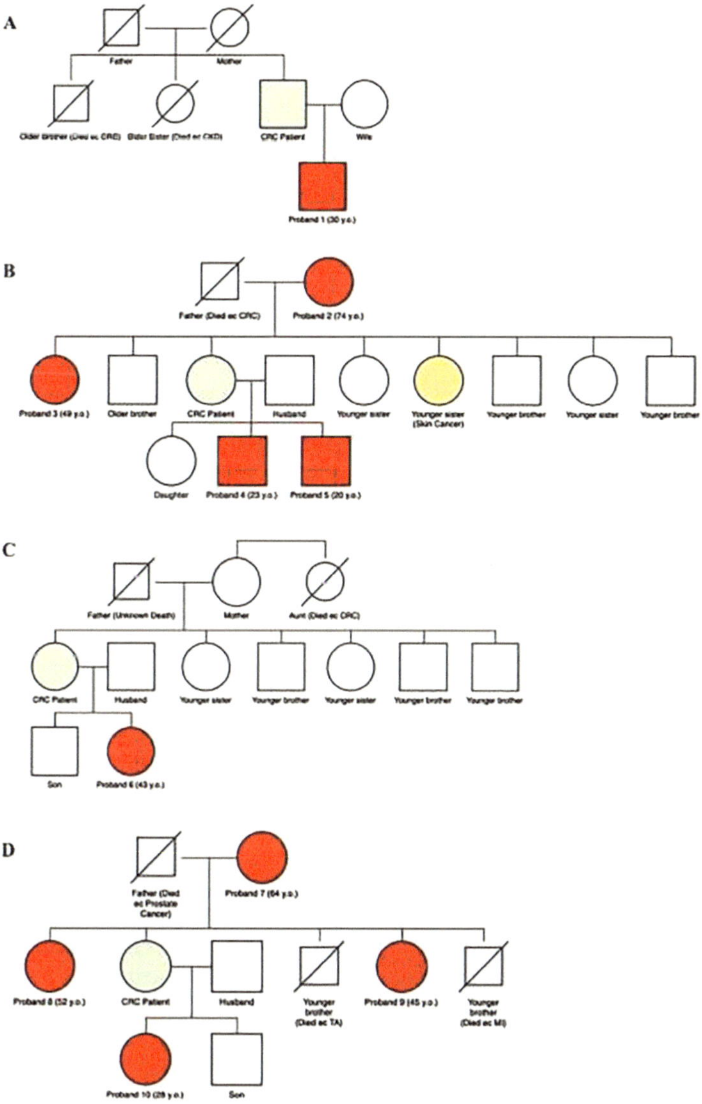


**Figure 2.**
*Hereditary CRC subjects' family pedigree. The pedigrees symbols correspond to the standardized human pedigree nomenclature [17], where a circle denoted a woman, a square for a man, a straight line for a relationship, and a diagonal strikethrough line for death. Colors also exhibited a similar trend. Red represented the proband, yellow for malignancy other than CRC, and light green for CRC subjects. A: Pedigree of 67-year-old man CRC, B: Pedigree of 44-years-old woman with early-onset CRC, C: Pedigree of elderly 62-years-old woman with CRC, D: Pedigree of 47-years-old woman with CRC.*

| Proband | Prior | Age (year) | *APC* (fc) | *MSH2* (fc) | *Y* | phi |
|---|---|---|---|---|---|---|
| 1 | 0.5 | 30 | 7290 | 9753 | 1 | 1.000 |
| 2 | 0.0 | 74 | 13,832 | 14,209 | 0 | 0.932 |
| 3 | 0.5 | 49 | 8727 | 9567 | 1 | 0.504 |
| 4 | 0.5 | 23 | 9757 | 10,320 | 1 | 0.505 |
| 5 | 0.5 | 20 | 14,524 | 13,073 | 0 | 0.500 |
| 6 | 0.5 | 43 | 11,676 | 10,673 | 1 | 0.500 |
| 7 | 0.0 | 64 | 14,020 | 13,653 | 0 | 0.500 |
| 8 | 0.5 | 52 | 6884 | 7073 | 1 | 0.500 |
| 9 | 0.5 | 45 | 14,341 | 13,295 | 0 | 0.500 |
| 10 | 0.5 | 28 | 14,609 | 13,426 | 0 | 0.500 |

**Table 6.**
*Proband Bayesian CRC risk estimation.*

**Table 6** described the process of CRC risk estimation among the probands. Bayesian analysis conditionally tweaked each proband's initial risk ('prior' column) with his or her age and gene expressions to yield adjusted odds ('phi' column) in developing or carrying CRC ('Y' column).

The combination of the pedigree in **Figure 2** and Bayesian estimation in **Table 6** gave rise to the complete story of CRC risks in 10 selected probands relative to the four subjects with CRC. A 67-year-old man with CRC of late-onset had a son with almost 100% CRC risk carrier or development (**Figure 2A**). Meanwhile, a 44-year-old woman with early-onset CRC and paternal death due to CRC had a mother with 93% unlikely, a son with 50.51% likely, and another son with 50% unlikely carrier or incidence of CRC (**Figure 2B**). An elderly woman with paternal death of unknown origin and CRC death of the aunt had an adult daughter with 50% of the CRC risk (**Figure 2C**). On the other side, a 47-year-old CRC diseased woman with a huge family and paternal death due to prostate cancer had a 50% likely risk on her sister, but three 50% unlikely probability on her mother, younger sister, and female offspring (**Figure 2D**).

## 4. Discussion

Several autosomal dominant diseases may appear to be majorly asymptomatic until adulthood or beyond puberty. Given the notion, sophisticated comprehension of the hereditary risk of neoplasm is critical. Direct genomic examination and molecular diagnosis of nucleic acids from the blood, tissue, or other bodily fluids have become more prominent as a screening and investigation standard [4].

DNA is the building block of every living cell in the world. In doing its job, DNA is often destringed into a single-stranded DNA (ssDNA) for DNA replication or mRNA transcription. The first case goes by attachment of complementary nucleic acid base pairs to the corresponding one in both the leading and lagging strand of the partly unzipped ssDNA by DNA polymerase. This created an exact copy of the source DNA. While for the latter case, the RNA polymerase enzyme works separately on the sense and antisense part of the ssDNA. Each strand of ssDNA produced a single mRNA, thus there will be two mRNAs for every transcription of a DNA. These mRNA then moved to the ribosome for translocations into amino acids and eventually proteins.

Cancerous cell arises due to mutations or faulty repair of the nucleic acid bases. Even one deletion, addition, or translocation of the bases drastically changes the transcribed mRNA, codon, and hence the protein. Commonly, uncontrolled proliferation of cells happened if the mistakes occurred on oncogenes or tumor suppressor genes. Constantly activated oncogenes or inhibited tumor suppressor genes direct the cell cycle to bypass checkpoints and not return to the resting phase.

Familial adenomatous polyposis and Lynch syndrome are the top two subtypes of hereditary colorectal cancer with the most incidence count. FAP is almost solely generated because of mistakes in the tumor suppressor gene of *APC*. Whilst a lot of MMR genes can be responsible for LS or HNPCC (e.g., *MLH1*, *MSH2*, *MSH6*, *PMS2*, and epithelial cell adhesion molecule) [24]. The present study accordingly chose APC and one of the repair genes to accommodate both hereditary subtypes of CRC. We inevitably selected *MSH2* among the other mismatch repair (MMR) genes due to the prevalence and missenses amount. Kim et al. stated that approximately 90% of the mutations in the MMR occurred in either *MLH1*, *MSH2*, or both genes [25]. Furthermore, a 13% increase of missenses (i.e., a type of gene mutation which renders genotypic reading and interpretation to be considerably harder) was measured on *MLH1* as opposed to *MSH2* [26]. *MLH* is also more often found in sporadic colon cancer [14]. Next, a three countries assessment on lynch syndrome also gathered that *MSH2* had a substantially higher 10-year-risk of severe adenoma ($\Delta$ = 10.1%) and tumor pathogenic variants (11.4 vs. 11.3%) over *MLH1* [27].

Observation among the CRC versus the control group displayed a lesser mean mRNA level of *APC* gene expression than the control group ($\Delta$ = 1683.06 fc, $p$ = 0.014). However, the reverse is true for the *MSH2* gene expression ($\Delta$ = 808.82 fc, $p$ = 0.465). The minute discrepancy can be because of many gene mutations also somatically involved *APC*. Engel et al. confirmed that from *MSH2*, *MSH6*, and *MLH1* tumor variants, somatic mutations of *APC* happened in 75, 100, and 11% cases [27].

The current study and its prior version in the Indonesian Medical Journal [28], swiftly acted on a novel proposition to determine *APC* and *MSH2* gene expressions cutoff for hereditary cancer classification. Looking over the percentile distribution of healthy controls, the measure of the first to fifth percentiles only had negligible insignificant differences. Following the wrapped Cauchy distribution of circular data techniques with M, D, and A statistics [29], we officialize the fifth percentile mark as the hereditary cutoff (*APC* = 12,195.80 fc and *MSH2* = 11,059.60 fc).

Interesting hereditary proportions had been exhibited by the 40 CRC subjects. Hereditary using only *APC* gene enlisted 50%, while *MSH2* gene gathered 32.5%, and both genes combination enticed 52.5% of the subjects. Only the one with *MSH2* cutoff showed akin prevalence to the 20–30% global familial CRC [30].

The brief, simple, yet informative presentation of the family medical history can be conveniently reflected through a visual pedigree. Taking accurate information on family history should be standard medical practice. The pedigree will subtly enhance oncology prevention, diagnosis, and treatment together with recent genomics advancements. Family history can substantially alter not only genetic testing results but also oncology prevention, including digestive cancer [16]. For instance, a study found that cancer occurrence in an individual relative to a bowel cancer was dependent on familial cancer prevalence, duration of onset, and closeness to the diseased [14]. The current study employs a family pedigree to analyze the health status of CRC patients' relatives. The diagram clearly outlines familial relationships; thus, it will be easier for recognition and interpretation of the inheritance patterns [16, 17].

Individualized medical care has been on the rise of attention. Instead of general medicine for a certain disease, therapeutic care needs to start keenly observing the patient and prescribing medicine that has been tailored for that particular individual [31]. One of the ways in achieving such goals is through personal genetic and biological factors consideration. Particularly, extra attention should be imposed on hereditary disorders like malignancy.

Relatives of a confirmed cancer patient may not experience any cancer-predisposing syndrome from a clinical viewpoint although having an increased risk of developing one. Decreased penetrance and onset age must be factored into consideration especially for autosomal dominant hereditary disorders like colorectal cancer [4, 7, 8, 32].

Investigation on the hereditary risk of diseases in relatives and families accordingly requires a lot of extra information and tests. The natural history of the disease firstly should be made clear (i.e., what is the diagnosis, how does it spread, how does it inherit, and what about its epidemiology). Next, focus on medical examinations or observations that include genetic data (i.e., marker, expression, mutation, and then clinical data). Lastly, we can examine the family pedigree and make objective evidence-based inferences for analysis.

Bayesian analysis of the pedigree from the familial and genetic factors displayed an interesting finding. It was obvious that the majority of probands on the next generation of the current CRC patients had a higher likely probability of CRC incidence or carrier. Probands below the current CRC patients' generation (i.e., the first, fourth, fifth, sixth, and tenth probands) on average had a 60.1% likelihood of CRC risk. Those in the same generation had 50.1% risk, while those above (i.e., the second and seventh probands) had 28.4% risk. These observations were moderately suitable with the autosomal dominant disease hereditary pattern.

The current study bridged the knowledge gaps of CRC hereditary cutoff, gene expressions risk of CRC, and Bayesian pedigree analysis. Nevertheless, some limitations are still presented. First, a small sample size increased the chance for a biased result. The sample selection in the study is also confined to a segmental niche of Indonesian urban citizens, hence the result may not be generalizable to rural or foreign populations. There were also no immunohistochemical or DNA sequencing mutation tests for objective comparisons. Lastly, the method of selecting cutoff from a fifth percentile has not broadly checked for its credibility yet.

## 5. Conclusion

The current study explored the relationship between *APC* and *MSH2* gene expressions to colorectal cancer risk assessment. Bayesian analysis computed that downregulation of the mRNA gene expression will induce a higher risk of developing or exacerbating CRCs. Yet only *APC* had significance while *MSH2* did not. Therefore, the study establishes the foundation of utilizing *APC* and *MSH2* gene expressions for CRCs risk indicators. Future novel or multiplicity studies should consider family pedigree as a part of CRC prevention strategy among the patient's relatives with expanded cohorts and sample pool, including more profound Bayesian analysis and application with other essential hereditary genes.

## Acknowledgements

The authors appreciate all of the participants in this study.

## Conflict of interest

The authors declare no conflict of interest. The authors acknowledge that although Vincent Tedjasaputra is currently an American Association for the Advancement of Science (AAAS) Science and Technology Policy Fellow; yet, the current paper is not affiliated with AAAS nor is the product of his position at the National Science Foundation.

## Declarations

Please note that an earlier version of this manuscript has been published in the Indonesian Biomedical Journal volume 12 on 2020 with a digital object identifier (DOI) of 10.18585/inabj.v12i4.1329 [28].

## Abbreviation summary

| | |
|---|---|
| APC | adenomatous polyposis coli |
| BMI | body mass index |
| bp | base pair |
| CKD | chronic kidney disease |
| CRC | colorectal cancer |
| DNA | deoxyribose nucleic acid |
| EDTA | ethylenediaminetetraacetic acid |
| EPCAM | epithelial cell adhesion molecule |
| FAP | familial adenomatous polyposis |
| fc | fold-change |
| GAPDH | glyceraldehyde 3-phosphate dehydrogenase |
| HNPCC | hereditary non-polyposis colon cancer |
| IRB | Institutional Review Board |
| LS | Lynch syndrome |
| MI | myocardial infarct |
| MLH | MutL homolog |
| MMR | mismatch repair |
| MSH | MutS homolog |
| PMS | post-meiotic segregation |
| RNA | ribose nucleic acid |
| RT-PCR | real-time-polymerase chain reaction |
| SD | standard deviation |
| ss | single-stranded |
| TA | traffic accident |
| TACSTD | tumor-associated calcium signal transducer |
| y.o. | years old |

## Author details

Tjahjadi Robert Tedjasaputra[1*], Mochammad Hatta[2], Muhammad Nasrum Massi[3], Rosdiana Natzir[4], Ilhamjaya Patellongi[5], Marcellus Simadibrata[6], Rina Masadah[7], Muhammad Luthfi Parewangi[8], Prihantono[9], Andi Asadul Islam[10], Agussalim Bukhari[11], Rinda Nariswati[12], Shirly Elisa Tedjasaputra[1], Vincent Tedjasaputra[13] and Jonathan Salim[14]

1 Internal Medicine Department, Tarakan General Hospital Jakarta, Jakarta, Indonesia

2 Faculty of Medicine, Molecular Biology and Immunology Laboratory, Universitas Hasanuddin, Makassar, Indonesia

3 Faculty of Medicine, Microbiology Department, Universitas Hasanuddin, Makassar, Indonesia

4 Faculty of Medicine, Biochemistry Department, Universitas Hasanuddin, Makassar, Indonesia

5 Faculty of Medicine, Physiology Department, Universitas Hasanuddin, Makassar, Indonesia

6 Faculty of Medicine, Internal Medicine Department, Universitas Indonesia, Jakarta, Indonesia

7 Faculty of Medicine, Pathology Anatomy Department, Universitas Hasanuddin, Makassar, Indonesia

8 Faculty of Medicine, Internal Medicine Department, Universitas Hasanuddin, Makassar, Indonesia

9 Faculty of Medicine, Oncology Surgery Department, Universitas Hasanuddin, Makassar, Indonesia

10 Faculty of Medicine, Neurosurgery Department, Universitas Hasanuddin, Makassar, Indonesia

11 Faculty of Medicine, Nutrition Department, Universitas Hasanuddin, Makassar, Indonesia

12 Statistic Department, School of Computer Science, Bina Nusantara University, Jakarta, Indonesia

13 American Association for Advancement of Science (AAAS), Science and Technology Policy Fellow, National Science Foundation (NSF), Washington, DC, USA

14 Faculty of Medicine, Pelita Harapan University, Tangerang Indonesia

*Address all correspondence to: rtedjasaputra@yahoo.com

## References

[1] Jasperson KW, Tuohy TM, Neklason DW, Burt RW. Hereditary and familial colon cancer. Gastroenterology. 2010;**138**(6):2044-2058. DOI: 10.1053/j.gastro.2010.01.054

[2] Giardiello FM. Hereditary colorectal cancer and polyp syndromes. In: Gearhart SL, Ahuja N, editors. Early Diagnosis and Treatment of Cancer Series: Colorectal Cancer. Philadelphia: Elsevier; 2010. pp. 21-30. DOI: 10.1016/B978-1-4160-4686-8.50008-7

[3] Nussbaum R, McInnes R, Willard H. Cancer genetics and genomic. In: Thompson & Thompson Genetics in Medicine. 8th ed. Philadelphia: Elsevier; 2015. pp. 309-332

[4] Nussbaum R, McInnes R, Willard H. Risk assessment and genetic counselling. In: Thompson & Thompson Genetics in Medicine. 8th ed. Philadelphia: Elsevier; 2015. pp. 333-348

[5] Tranø G, Sjursen W, Wasmuth HH, Hofsli E, Vatten LJ. Performance of clinical guidelines compared with molecular tumour screening methods in identifying possible Lynch syndrome among colorectal cancer patients: A Norwegian population-based study. British Journal of Cancer. 2010;**102**(3):482-488. DOI: 10.1038/sj.bjc.6605509

[6] Turnpenny P, Ellard S, Cleaver R. Patterns of inheritance. In: Emery's Elements of Medical Genetics and Genomics. 16th ed. Philadelphia: Elsevier; 2020. pp. 66-82

[7] Turnpenny P, Ellard S, Cleaver R. Risk calculation. In: Emery's Elements of Medical Genetics and Genomics. 16th ed. Philadelphia: Elsevier; 2020. pp. 94-101

[8] Nussbaum R, McInnes R, Willard H. Patterns of single-gene inheritance. In: Thompson & Thompson Genetics in Medicine. 8th ed. Philadelphia: Elsevier; 2015. pp. 107-132

[9] Turnpenny P, Ellard S, Cleaver R. The cellular and molecular basis of inheritance. In: Emery's Elements of Medical Genetics and Genomics. 16th ed. Philadelphia: Elsevier; 2020. pp. 9-23

[10] Buckingham L. Molecular basis of inherited disease chromosomal abnormality. In: Buckingham L, Flaws ML, editors. Molecular diagnostics. Fundamentals, Methods, & Clinical Applications. Philadelphia: F. A. Davis Company; 2007. pp. 311-326

[11] Vogelstein B, Papadopoulos N, Velculescu VE, Zhou S, Diaz LA, Kinzler KW. Cancer genome landscapes. Science (80). 2013;**339**(6127):1546-1558. DOI: 10.1126/science.1235122

[12] Teama S. DNA polymorphisms: DNA-based molecular markers and their application in medicine. In: Yamin Liu, editor. Genetic Diversity and Disease Susceptibility. London: Intech; 2018. DOI: 10.5772/intechopen.79517

[13] Li M, Wang IX, Li Y, Bruzel A, Richards AL, Toung JM, et al. Widespread RNA and DNA sequence differences in the human transcriptome. Science (80). 2011;**333**(6038):53-58. DOI: 10.1126/science.1207018

[14] Turnpenny P, Ellard S, Cleaver R. The genetics of cancer and cancer genetics. In: Emery's Elements of Medical Genetics and Genomics. 16th ed. Philadelphia: Elsevier; 2020. pp. 117-199

[15] Guttmacher AE, Collins FS, Carmona RH. The family history—More

important than ever. The New England Journal of Medicine. 2004;**351**(22): 2333-2336. DOI: 10.1056/NEJMsb042979

[16] Brock J-AK, Allen VM, Kieser K, Langlois S. Family history screening: Use of the three generation pedigree in clinical practice. Journal of Obstetrics and Gynaecology Canada. 2010;**32**(7):663-672. DOI: 10.1016/S1701-2163(16)34570-4

[17] Bennett RL, French KS, Resta RG, Doyle DL. Standardized human pedigree nomenclature: Update and assessment of the recommendations of the National Society of Genetic Counselors. Journal of Genetic Counseling. 2008;**17**(5):424-433. DOI: 10.1007/s10897-008-9169-9

[18] Takara Bio USA Inc. Mini & midi spin kits for isolation of total RNA from blood—NucleoSpin RNA Blood & NucleoSpin RNA Blood Midi [Internet]. United States of America; 2017

[19] Zhou Z, Liu H, Wang C, Lu Q, Huang Q, Zheng C, et al. Long non-coding RNAs as novel expression signatures modulate DNA damage and repair in cadmium toxicology. Scientific Reports. 2015;**5**(1):15293. DOI: 10.1038/srep15293

[20] Hatta M, Surachmanto EE, Islam AA, Wahid S. Expression of mRNA IL-17F and sIL-17F in atopic asthma patients. BMC Research Notes. 2017;**10**(1):202. DOI: 10.1186/s13104-017-2517-9

[21] Sirait R, Hatta M, Ramli M, Islam A, Arief S. Systemic lidocaine inhibits high-mobility group box 1 messenger ribonucleic acid expression and protein in BALB/c mice after closed fracture musculoskeletal injury. Saudi Journal of Anaesthesia. 2018;**12**(3):395. DOI: 10.4103/sja.SJA_685_17

[22] Tambaip T, Br Karo M, Hatta M, Dwiyanti R, Natzir R, Nasrum Mas M, et al. Immunomodulatory effect of orally red fruit (*Pandanus conoideus*) extract on the expression of CC chemokine receptor 5 mRNA in HIV patients with antiretroviral therapy. Research Journal of Immunology. 2018;**11**(1):15-21. DOI: 10.3923/rji.2018.15.21

[23] University of Twente. PCR Primer Information. 2017. Available from: https://www.utwente.nl/en/tnw/dbe/research/pcp-primers/ [Cited: 29 December 2019]

[24] Rashid MU, Naeemi H, Muhammad N, Loya A, Lubiński J, Jakubowska A, et al. Prevalence and spectrum of MLH1, MSH2, and MSH6 pathogenic germline variants in Pakistani colorectal cancer patients. Hered Cancer Clinical Practices. 2019;**17**(1):29. DOI: 10.1186/s13053-019-0128-2

[25] Kim Y-M, Choe C-G, KimCho S-M, Jung I-H, Chang W-Y, Cho M-J. Three novel germline mutations in MLH1 and MSH2 in families with Lynch syndrome living on Jeju Island, Korea. BMB Reports. 2010;**43**(10):693-697. DOI: 10.5483/BMBRep.2010.43.10.693

[26] Lynch HT, de la Chapelle A. Hereditary colorectal cancer. The New England Journal of Medicine. 2003;**348**(10):919-932. DOI: 10.1056/NEJMra012242

[27] Engel C, Ahadova A, Seppälä TT, Aretz S, Bigirwamungu-Bargeman M, Bläker H, et al. Associations of pathogenic variants in MLH1, MSH2, and MSH6 with risk of colorectal adenomas and tumors and with somatic mutations in patients with lynch syndrome. Gastroenterology. 2020;**158**(5):1326-1333. DOI: 10.1053/j.gastro.2019.12.032

[28] Tedjasaputra TR, Hatta M, Massi MN, Natzir R, Patellongi I, Simadibrata M, et al. Risk assessment

in hereditary colorectal cancer family by using APC and MSH2 mRNA gene expression and Bayesian analysis. Indonesia Biomedical Journal. 2020;**12**(4):368-375. DOI: 10.18585/inabj.v12i4.1329

[29] Abuzaid AH, El-hanjouri MM, Kulab MM. On discordance tests for the wrapped cauchy distribution. Open Journal of Statistics. 2015;**5**(4):245-253. DOI: 10.4236/ojs.2015.54026

[30] Brosens LAA, Offerhaus GJA, Giardiello FM. Hereditary colorectal cancer. The Surgical Clinics of North America. 2015;**95**(5):1067-1080. DOI: 10.1016/j.suc.2015.05.004

[31] Meiliana A, Dewi NM, Wijaya A. Personalized medicine: The future of health care. Indonesia Biomedical Journal. 2016;**8**(3):127. DOI: 10.18585/inabj.v8i3.271

[32] Chin L, Andersen JN, Futreal PA. Cancer genomics: From discovery science to personalized medicine. Nature Medicine. 2011;**17**(3):297-303. DOI: 10.1038/nm.2323

Chapter 7

# Circular DNA: How Circular DNA Assists Cancer Roll with Therapeutic Punches

*Parvaiz Yousuf*


## Abstract

DNA within cells is either present in the form of long strands as in eukaryotes or circular shapes in Yeast plasmids, mitochondrial DNA, and double minutes in tumor cells. Apart from them, ribosomal or telomeric DNA has been found to produce specialized forms of extrachromosomal circular DNA (eccDNA). eccDNA was discovered in both normal and cancer cells in recent times, indicating a much more significant role. The eccDNA has been found to promote tumor proliferation, survival, and aggressiveness in almost half of all cancers by increasing oncogene copy numbers. This chapter will discuss the biogenesis and function of eccDNA and how it promotes tumor adaption under changing microtumour environmental conditions, as in the case of drugs.


**Keywords:** DNA, DNA, circular, neoplasms, antineoplastic protocols, oncogenes

## 1. Introduction

DNA is the basic genetic material present in most living species. It was first discovered by Friedrich Miescher in 1869 [1]. This genetic material exists in the form of chromosomes that consist of linearized double-stranded DNA fused with histone proteins within the nucleus and store most of the genetic information of an organism [2]. However, DNA or genes are not confined to the nucleus only as they can also be found in other extra-nuclear structures. Eukaryotes have circular DNA present within the mitochondria and chloroplast and are structurally similar to the bacterial genome [3, 4]. In addition, other forms of circular DNA also exist within the cytoplasm and nucleus [5]. This DNA has been named eccDNA, and its size varies from a few base pairs to millions of base pairs. On the basis of size and sequence, this form of DNA has further been categorized into 100 bp to 10 kb long small polydispersed DNA (spcDNA), 100 to 400 base pair long micro-DNA, millions of base pairs long extrachromosomal DNA, and telomeric circles or t-circles consisting of base pairs that are multiplies of 738. In addition to eukaryotes, viruses and bacteria also possess circular DNA in different forms with varying lengths [6–15], which are summarized in **Table 1**. In recent years, a lot of studies have been done on this form of DNA that has improved our knowledge. Their biological, physiological properties, as well as



IntechOpen

| Type of circular DNA | Size | Functions | Refs |
|---|---|---|---|
| *Viruses* | | | |
| ssDNAs | 2–6 kb | Role in replication | [6] |
| Retroviral DNA | 7–12 kb | Role in replication | [7] |
| dsDNA | 6–375 kb | Role in replication | [8] |
| *Bacteria* | | | |
| Plasmids | 30–2430 kb | Role in reproduction, drug resistance etc. | [9] |
| *Eukaryotes* | | | |
| Mitochondrial DNA | 16 kb | Maintains mitochondrial function | [10] |
| MicroDNA | 100–400 kb | Produces miRNAs | [11] |
| Double minute | 100 kb–3 Mb | Extrachromosomal gene amplification | [12, 13] |
| Telomeric circle | Integral multiplies | Restore telomere length of 738 bp | [14] |
| S small polydispersed circular DNA | 100 bp–10 kb | Enhances genomic stability | [15] |

**Table 1.**
*Circular DNA is present in several forms among viruses, bacteria and eukaryotes with different size ranges and functions.*

their functions, are becoming more noticed. This has led to the discovery of more previously unknown forms of circular DNA [16]. Some researchers believed that DNA exists in rings within the cytoplasm of higher organisms, which was later verified with the eccDNA discovery [17]. However, scientists observed this circular DNA in many other organisms before they were seen present in many heterogeneous cancer cells. For instance, Alix and Yasuo had observed the different lengths of eecDNA in sperm of boar observed under an electron microscope [5]. In many other eukaryotic species, such as humans, Yeast, hamsters, mice, and Drosophila, this form of DNA was observed [18–23]. At the same time, the size of eccDNA present in normal eukaryotic cells has been found to be very small and is usually less than 500 base pairs [20, 22–25]. However, eccDNA of a much bigger size has also been discovered in certain types of tumor cells. After the tumors were surgically removed and mitotic abnormalities were investigated after being stained by Cox, Spriggs, and acetic orcein, small chromatin bodies of the size of an intact chromosome were observed [26]. The name given to these chromosomes was double minutes (DMs) as they were present in pairs. Studies by Sprigg later concluded that such chromatin bodies were present more in malignant childhood brain tumors [27]. These DMs were also spotted in HeLa cells using the buoyant density method [28]. Thus, it is clear that the eccDNA was more commonly found in genetically unstable cells, such as tumor cells, and was not as common in normal cells [29, 30]. Although this DNA has been found to be homologous to the genomic DNA, it is clear that the presence of eccDNA means genome instability. This is why most of the tumors have exceptionally high levels of eccDNA, and the oncogenes usually occur in this genome only. Much research has been conducted to find out how this DNA carries driver oncogenes and contributes to tumor resistance and heterogeneity. The levels of eccDNA are much more common in tumor cells than was earlier thought. It makes tumors highly resistant to the therapeutic drugs by increasing the oncogene copy number and heterogeneity. This chapter explains in detail the eccDNA discovery in tumors and how it will shape the therapeutic operations for the treatment of varied cancer types.

## 2. eccDNA: from hypothetical existence to role in cancer in eukaryotes

The eccDNA was once thought to play some secondary roles in both prokaryotes and eukaryotes. However, with the advent of technology, it became clear that eccDNA plays more pivotal functions in eukaryotes than ever thought. It is estimated that over half of all tumor types have eccDNA, enabling them to develop resistance against different therapeutic drugs.

### 2.1 Historical perspective

The eccDNA was discovered by Yasuo Hoota and Alix Bassel [5]. They were actually investigating a theory proposed by Franklin Stahl in 1964 that higher organisms' chromosomes consist of DNA rings [5]. Their experiments lead to the discovery of DNA circles of varying sizes that range from 100s to 1000s of bp's resembling DNA of higher organisms. Double minutes (DM'S) was the name given to these large circles of DNA. Later on, scientists began investigating the presence of eecDNA in other cell types. For instance, another group of researchers discovered the DMs in human tumor cells by preparing karyotypes and utilizing purification by CsCl gradience [26, 28]. The DNA from many other organisms was studied using EM imaging techniques and CsCl gradient purification [22, 31–36]. The researcher used the technique of Southern Blotting to determine the homology between genomic DNA and eccDNA. Most of the eccDNA observed was less than 500 bps in size and is called poly-disperse circular DNA (spcDNA) [24, 37]. Moreover, most of the DNA obtained came from repetitive sequences of DNA, while some of the spcDNA molecules fused with particular sequences. Even some researchers observed that 9–11 base pair long direct repeats flank few non-repetitive spcDNA sequences on both sides [19–21]. This indicated that DNA circles were formed by certain DNA repair pathways, such as microhomology-mediated end joining or homologous recombination, which mediate the joining of ends in between small base repeats. However, many later studies proved that no repetitive sequences are needed to mediate this end-to-end joining. This occurred when eccDNA with specific sequences was isolated and sequenced with no repetitive areas within or flank the DNA [38]. Around the same time period, another group of researchers utilized exonuclease III for quantification of eccDNA and concluded that there occur varying levels of eccDNA among mice tissues [39]. As far as the formation of eccDNA is concerned, many groups used techniques studying eccDNA from repetitive sequences to determine their formation. 70 times increase in eccDNA formation occurred in murine cells when Cycloheximide (a protein synthesis inhibitor) was used. The same results were obtained with many other chemicals, such as hydroxyurea (DNA replication inhibitor, and 7,1-dimethylbenz[a]anthracene (a carcinogen) [20]. Similarly, a higher concentration of longer eccDNA molecules was observed in cells obtained from patients suffering from Fanconi Anemia (wherein particular DNA repair pathway errors occur) [30]. Afterwards, smaller eccDNA molecules were explored upon the usage of 2-dimensional gel electrophoresis, which showed that eccDNA levels are increased by carcinogens, and different stages of development show varied eecDNA formation in flies and frogs. It has been found that even the foreign DNA can result in the formation of eccDNA and DNA sequence organization within tandem repeats prepared eccDNA preparation from the DNA [40–42]. In a nutshell, it suggests that the formation of eccDNA depends on DNA organization, DNA damage repair, and the sequence of DNA. All this indicates how eecDNA has the tendency to code for driver oncogenes as eccDNA are amplified in terms of oncogenes leading to drug resistance in most cancers [43–45].

### 2.2 Recent advances in eccDNA

The eecDNA is not well understood, and researchers have been trying to understand it better. Human cancer and mouse tissue cells were lyzed, and eccDNA was isolated to recently sequence the entire length of an eccDNA compliment. This was done using a paired-end high sequencing technology that allowed the characterization of this form of DNA [46]. The junctional sequences of this eccDNA were identified through high-end sequencing technology on exonuclease-resistant eecDNA that was amplified using a rolling circle mechanism [46]. After these studies, it was found that all human cancer and mouse tissue cells showed consistent patterns with specific eccDNA sizes peaking around 180 and 380 base pairs. Around 5% of these molecules extended in the range of 2–3 kilobase pairs. The term given to this DNA was microDNA. However, the study may have resulted in the under-representation of long eccDNA molecules as small DNA circles are amplified more efficiently than large-sized circles. Moreover, most of the eccDNA circles were smaller, as represented by electron microscopy [47]. After mapping this eccDNA, many bases extended to over 100–1000 specific sites with the genome. It also showed enrichment in some particular regions, transcriptionally active chromatin, hotspots that include CpG regions and UTR regions, and DNA patches with high GC content (around 60%) [46]. Most microDNA is flanked by the genomic DNA containing 2–15 base pair repeats, suggesting a micro-homology mode generates DNA circles [46]. The sites where eccDNA is formed might be associated with cell lineage, which is indicated by weak clustering of eccDNA in the ovarian and prostate cell lines. Other studies claimed that if MSH3, which encodes a peptide in the DNA mismatch repair pathway, is deleted, it results in an 80% decrease in the eccDNA levels [47]. Moreover, there is no clear proof of whether this small eccDNA replicates or not. However, there have been rough estimates of electron microscopy about the eccDNA abundance prepared from a specific number of cells, indicating that 125–200 circles of eccDNA exist per DT40 cell [47]. Similarly, another study was performed in *Saccharomyces cerevisiae,* where researchers found 1/4th of the Saccharomyces being covered with about 2000 DNA circles. Moreover, this study ignored smaller eccDNA (less than 1-kilo base pair in size), and there was no dependency on the junctional sequence identification to label a particular DNA sequence as eccDNA. Therefore, the eccDNA was 1–38 kilobase pairs long, which were significantly enriched with circles from repeated genome parts, such as gene duplications, ribosomal DNA circles, and transposons. This suggests that circles were formed in a homologous recombination form. Furthermore, specific sequences were the precursors of around 60% eccDNA, and seven base pair direct repeats were revealed by over 90% of the genomic sites. This means that DNA circles are in a microhomology-directed mechanism.

The mechanism responsible for eccDNA formation is yet to be understood clearly. However, researchers have found that tandemly repeated genomic sequences are present in eccDNA [30, 48, 49]. It indicates that eccDNA formation primarily occurs from such tandemly repetitive DNA sequences [48]. At the same time, nonrepetitive

DNA also results in eccDNA formation. For instance, a group of researchers isolated eccDNA from HeLa S3 cells and found the presence of eccDNA [28]. May it be coding or non-coding regions, eccDNAs originate from both. Many studies concluded DMs bear drug-resistant oncogenes and another oncogene [50]. Different researchers have proposed four distinct models of how eccDNA formation occurs. The first one is the translocation-deletion-amplification model, in which genes' rearrangements occur near the chromosome's translocation site [51]. The fragments closer to the translocation breakpoints can be deleted, amplified, and circularized, which results in eccDNA formation [52, 53]. The second model-Chromothripsis model, explains that multiple acentric DNA fragments can form due to chromosome shattering, with some of the fragments self-ligating to form eccDNAs [54, 55]. The third model is the breakage-fusion-bridge (BFB) model [56]. Losing a telomere by a chromosome is what initiates the BFB cycle. Two chromatids are formed during anaphase after chromosome duplicates. Afterwards, the fusion between broken chromosome ends (chromatids) occurs, which results in the formation of a dicentric chromosome [57]. A bridge is formed between these two chromatids during anaphase due to the presence of two centromeres that are disrupted when two centromeres are directed to opposite poles of the spindle: The chromosome breakage and uneven distribution of genetic material results in the centromeres being segregated into daughter cells [58]. Thus, one of the daughter cells has a chromosome with an inverted repetitive base sequence on its ends, while the other has a chromosome with deletion at the ends. When DNA replication continues in the next cycle, the fusion of sister chromatids occurs once again, resulting in the repetition of the BFB cycle. All the events result in DNA sequence amplification at the telomeres that eventually loops out to form eccDNAs [59]. The fourth model is the episome model, which explains that excision of small circular DNA results in episome formation, which eventually results in recombination or over-replication, resulting in eccDNA formation [34, 60]. It is well known that eccDNAs can lead to mutations, amplifications, translocations, deletions, etc. In human somatic cells, certain loci that include HLA, DAZ4, and KIR have been found to be more prone to circular product formation, resulting in chromosomal deletions [61]. Similarly, in a study performed on Yeast, identification of 1756 eccDNAs was made, which covered 23% of the total Yeast genome [62]. This means it is highly likely for an oncogene to be present on eccDNA. Similarly, the presence or absence of DMs is a vital factor to consider as it relates to the clinical outcomes of the patients. For example, some oncogenes, such as oncogene Sei-1, induce DM formation [63]. Similarly, another study found that oncogene Met's higher amplification and expression occurs when present on DMs. The promotion of Sei-1 induced DM generation results because of this Met signaling cascade pathway. Moreover, in the patients' ovarian cancer (OC) cells, a greater copy number amplification of eukaryotic initiation factor 5A2 (EIF5A2) was found by way of DMs, indicating the role of DMs in carrying and sustaining oncogenes [64].

Recently, a group of researchers studied human cell lines and *Caenorhabditis elegans* by using a mechanism that relied on Cesium Chloride- Ethidium Bromide (density gradient centrifugation) and high-throughput sequencing and tagmentation [65]. They reported circles that mapped on non-coding and coding genomic regions. The protein-coding regions of the DNA encoding titin and mucin lead to the appearance of eccDNA in these cases. The study concluded that the eccDNA promotes and interferes with the particular exon transcription, thereby leading to the expression of various isoforms of a genetic code [46]. In cancerous cells, the eccDNA has been found to play complex roles to promote tumourigenesis. The changes in DMs, eccDNAs between normal and cancerous cells have been quantified by fluorescent microscopy [12, 66]. The amplification of myc and EGFR genes in tumor cells occurred through a few passages by eccDNA formation [66]. Similarly, another group of researchers combined fluorescence imaging with software analyzing images for quantification of certain oncogene copies. There was a 40% amplification of EGFR and MYC gene in the examined human cancer cells, whereas normal cells showed no enrichment [12]. What surprised researchers were that more oncogene amplification occurred through a mechanism involving eccDNA formation rather than a mechanism involving chromosomal amplification [12]. The detection of eccDNA within the normal tissues may be primarily because of having no procedures helping circle enrichment and not enough dye binds to small microDNA (eccDNA) for microscopic detection. All of these studies prove solid evidence that tumor heterogeneity is a result of eccDNA, wherein they also assist cancer cells in evolving through an increase in oncogene copy number [47]. The validation of these studies was done by another group of researchers that found MET (an oncogene) in glioblastoma cells showed amplification on eccDNA as was indicated by FISH [67].

## 3. Extrachromosomal oncogene amplification drives tumor evolution and genetic heterogeneity

Human tumors are highly heterogeneous, and they evolve and adapt in quickly changing microenvironments from individual cells to a mass of genetically heterogeneous cells. Only those cells are selected by the Darwinian selection, which adapts quickly to their environments. Tumor heterogeneity offers a mutation pool from which tumor-friendly mutations are selected [68–72]. The progression of neoplasm and resistance to therapeutic drugs is driven when mutations are passed on to daughter cells after cells have acquired the mutations enhancing fitness. For instance, one of the usual mutations of cancer cells is oncogene amplification, which is either present on chromosomal DNA or eccDNA parts that include DMs too [60, 73, 74]. Compared to the chromosomal DNA, the eccDNA does not have high stability and segregates to daughter cells in an unequal fashion (**Figure 1**). As per the Mitelman database, 1.4% of cancers possess DMs, with neuroblastoma showing a maximum of 31.7% DMs [75]. However, no accurate quantification of the eccDNA has been done in the cancer cells, as there has been no systemic examination of oncogenes present in this eccDNA. At the same time, how eccDNA impacts tumor cell evolution is yet to be understood well. Although we can sequence DNA in an unbiased manner to analyze the cancer genes, the spatial resolution of amplicons is not possible yet to determine the specific regions of chromosomes of eccDNA. Moreover, DNA circularity can potentially be inferred by using bioinformatic analysis [76], but eccDNA amplicons may show variations from cell to cell. This means there has been a great underestimation of oncogene amplification related to eccDNA. Although by cytogenetically analyzing the cancer cell metaphases, the localization of amplicons can be done, there is always some bias associated with this technique. Recently, a group of researchers quantified the eccDNA spectrum in human tumor cells and systemically interrogated the contents by integrating the Whole Genome Sequencing (WGS) of 117 human tumor cell lines, varied tumor tissues from various cancer types, and cancer cell cultures derived from patients [12]. The researchers analyzed 2049 metaphases from around 72 samples of cancer cells bioinformatically and cytogenetically. In addition, they analyzed a total of 233 metaphases from eight normal tissue cultures and 290 metaphases from 10 immortal

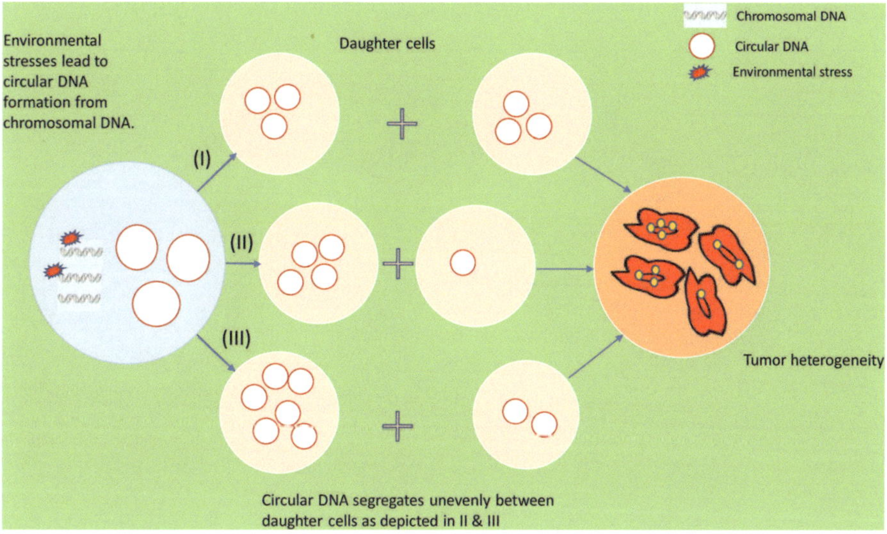


**Figure 1.**
*Environmental stresses lead to the formation of circular DNA from chromosomal DNA, which can be (I) equally distributed or (II and III) unequally distributed as in most of the cases, thereby increasing heterogeneity of oncogenes when present.*

cancer cell cultures with a sum of 2572 metaphases. The eccDNA is detected by the DAPI, 4′, 6-diamidino-2-phenylindole (a fluorescent dye), which FISH and genomic DNA probes confirm. As expected, they found that eccDNA is present abundantly in cancer cell samples and was seldom present in normal cells. This shows eccDNA may have a closer association with the tumor cells than earlier thought. Moreover, almost 30% of the eccDNAs were found to be paired DMs [12]. However, it is necessary to mention that different tumors showed different levels of eccDNA, and the levels were much higher in the cultures obtained from patients. In two of 20 metaphases, the conservative metric of two eccDNAs was used, and approximately 40% of cancer cell lines and 90% of brain tumor models from patients were found to possess eccDNA [12]. There is no association between levels of eccDNA and either treated versus untreated samples, metastatic versus primary status, tumors not irradiated versus tumors that were irradiated. Moreover, with respect to the size of samples taken, it is tougher to determine the effect of different therapies on eccDNA levels undertaking the various treatment options available. There is a great variation of eccDNA number within this tumor cell culture between cells. Thus, it is confirmed that the levels of eccDNA in cancer cells are very common; however, there may be variations from cell to cell. At the same time, the eccDNA levels in normal tissues are quite rare. Using Whole Genome Sequencing, focal amplifications were revealed from cancer cell lines of different types that included amplified oncogenes defined earlier from 13 varied types of tumors [77]. Surprisingly, most cancer oncogenes are present on eccDNA only and other Homologous Staining regions of the chromosomes. Moreover, mRNA transcripts are high expression levels within the oncogenes present on eccDNA. At the same time, the diversity in the copy number of oncogenes present on eccDNA is much greater than the diversity of the copy number of genes presents on other chromosomes (**Figure 2**) [12].

Researchers have studied the origin of both intra- and extra-chromosomal structures and try to determine whether they originate from the same or different

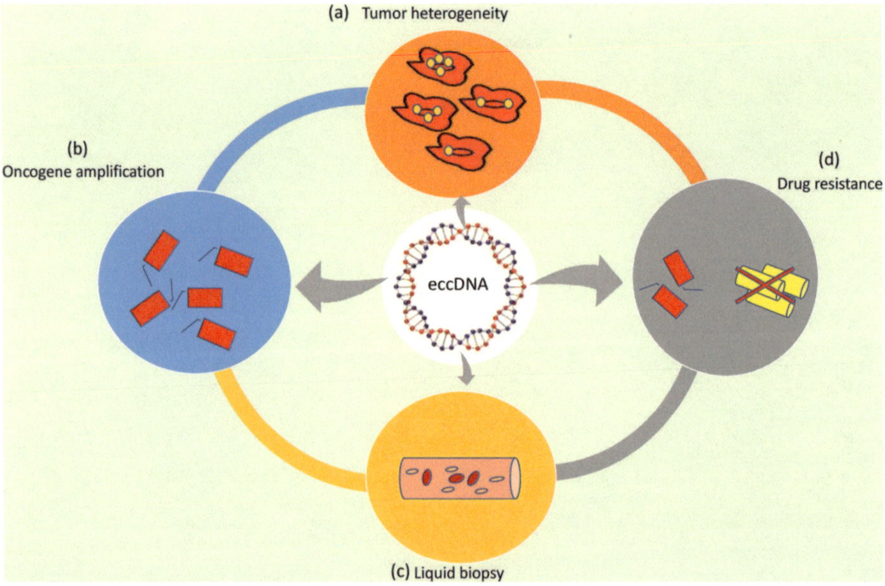


**Figure 2.**
*EccDNA plays complex roles in a TME which include (a) tumor heterogeneity: Tumor cells shown in red receive an unequal number of circular DNA molecules shown as yellow dots. (b) Oncogene amplification: Increase in a number of a particular oncogene as depicted by red-tailed blocks (c) liquid biopsy: A test used to detect cancer by measuring circulating eccDNA levels as shown by dark-red ovals (d) inducing drug resistance: Oncogenes develop resistance to drugs depicted as yellow cylinders.*

precursors. The relationship that exists between amplicon and sub-nuclear regions was understood by taking advantage of GBM39 subclone cells, which occur spontaneously and in which EGFRvIIIs high copies shift from eccDNA to Homogeneous Staining Regions [12]. The eccDNA amplicon on separate GBM39 replicates showed a circle-shaped structure with about 1.29 Megabases that contained 1 EGFRvIII copy. Surprisingly, the GBM39 subclone, which harbored EGFRvIII on homogeneous staining regions, showed a similar structure with tandem duplications containing multiple EGFRvIII copies. This indicates that EGFRvIII, which contained eeCDNA structures, integrated to form Heterogeneous Staining Regions. On this eeCDNA, the reversible loss of EGFRvIII leads to resistance in GBM39 cells to the inhibitors of EGFR tyrosine kinase [78]. The EGFRvIII amplicon that contains eccDNA is conserved in the naïve cells as indicated by the structural analysis of these structures, which indicates that eccDNA maintains its primary structural properties by relocating to HSRs on chromosomes [13, 60].

Now, whether is the localization of eccDNA is beneficial or not is yet to be studied. However, a hypothesis about the same exists, which postulates that eccDNA amplification may be responsible for helping an oncogene get a high copy number. This way, tumor genes are able to replicate and start forming products necessary for the overall survival and growth of a tumor. This hypothesis may be true as eccDNA containing oncogenes is unequally segregated to the tumor cells due to intrachromosomal amplification [79]. Moreover, experiments have been carried out using a branching process (Simplified Galton-Watson) to explain how tumors evolve. In these experiments, either replication or death of a cell in the current generation occurs to build the upcoming generation. Considering some assumptions, they found that independent segregation of the eccDNA copies into the two daughter cells occurs during the process of cell division. Moreover, the same study confirmed that oncogene favors the development and replication of an oncogene in a better way. To strengthen and prove this point, the researchers have found a greater copy number of oncogenes c-MYC and EGFR in the eccDNA rather than in the chromosomes. Moreover, the intra-chromosomal amplification of an oncogene results in the stabilization of heterogeneity of a tumor at a lower level. In comparison, the eccDNA is unequally segregated, resulting in greater tumor heterogeneity and maintaining it [80, 81]. Also, the copy number of an oncogene may increase the tumor heterogeneity but is likely to do so quite slowly if present on a chromosome than if the same oncogene is present on an eccDNA molecule. Moreover, there is strong evidence that it is quite difficult to cure genetically heterogeneous tumors [80]. The tumor cells have the ability to maintain the oncogene transcriptional level and copy number variability from cell to cell, which results in resistance to drugs and the progression of tumors. The worst thing about eccDNA is that the oncogene amplification in eccDNA enables the tumor adaptability more effectively, helping it survive tough microenvironment conditions. This is further done because of an increase in the chances that a particular cell population expresses a specific oncogene at such a level that considerably increases the survival and proliferation of a tumor [60, 78, 82–85]. This worsens the situation even more as it becomes immensely difficult to treat such tumors after they become progressively aggressive. Thus, the extra and intrachromosomal amplification mechanism varies greatly, leading to high copy number heterogeneity and greater amplicons copy number. It means that even if oncogenes present on eccDNA confer a little selection advantage increase, it will still lead to a greater advantage of fitness for a tumor. Thus, as far as the evolution of a tumor is concerned, there is certainly a great role for oncogene present on eccDNA to play as they are linked to tumor heterogeneity and greater survival. This does not happen when the same gene (even if) is present on a chromosome. Thus, it is of immense importance and vitality to understand how a tumor evolves on a molecular level and how oncogene amplification occurs in eccDNA. This will help the scientific community to effectively treat tumors by either preventing their progression or successfully eradicating them.

When it comes to detecting eccDNA, the circulating levels of this DNA aid in the noninvasive diagnosis of tumors, thereby helping manage them. This becomes possible because tumors and normal tissues have the property of releasing eccDNA into circulation. The eccDNA has been observed in the mammalian tissue nuclei and other cell lines. However, researchers are demonstrating that eccDNA with unique mapping regions are detectable in serum and plasma obtained from both humans and mice [85]. Moreover, the eccDNA obtained from the serum and plasma has longer lengths with over 250 base pairs than the one found in cells (around 150 base pairs). Researchers have detected human microDNA in mouse circulation after transferring human cancer lines. Moreover, when microDNA from normal cells and tumor cells was compared, it revealed that longer micro-DNA occurred in the tumor cells. Moreover, researchers even collected the micrDNA samples from cancer patients before and after the surgery and found that longer and higher concentrations of microDNA are released into circulation when the tumor is there and shorter when the tumor is excised. This indicates that circular DNA is not confined to cells (cytoplasm) only but can be thought of as everywhere in the presence of a tumor [86].

### 3.1 Functional importance of eccDNA with emphasis on tumors

The eccDNA performs a varied number of functions within and around the cells. In tumor presence, a person shows a higher concentration of circular DNA in and around the cells, indicating an ever-thought greater role. Although insufficient research has been performed on the eccDNA and its functions in tumors, they certainly have been found to contribute to genetic heterogeneity in tumor cells. They are responsible for oncogene amplification, a hallmark of many cancer types, thereby inducing drug resistance. Thus, a particular tumor is more likely to express an oncogene on the eccDNA, providing it with a suitable environment for the proliferation, progression, and metastasis of a tumor. Apart from this, numerous theoretical functions of eccDNA have been reported, including the gene dosage compensation, heterogeneity among cells, transcription factor sponging. Moreover, they have a predominant role in producing a mutational pool of DNA that helps tumors evolve, a role in intercellular communication, aging, stimulating certain innate immune pathways, and uses in the technique of liquid biopsy. Numerous researchers have studied the role of eccDNA in tumors and unequivocally revealed the role of eccDNA in tumors and inducing drug resistance. For instance, several drug-resistant genes and oncogenes are carried on DMs, which leads to the advancement of cancer by the phenomenon of amplification of genes [13, 87–89]. With more advanced research being done in the field, it was revealed that it is much more common for eccDNA to mediate amplification of genes that thought earlier. Thus, the presence or absence of eccDNA is vital for the evolution and heterogeneity of tumors. For instance, around 40% of the cancer cell lines and around 90% of brain tumor cell lines from patients showed eccDNA presence [41]. Apart from this, researchers have provided experimental and mathematical evidence suggesting that heterogeneity among tumors and amplification of driver oncogenes are much greater upon amplification on eccDNA than when the same happens on chromosomal loci. Furthermore, an immense proliferation advantage is acquired when eccDNA is distributed randomly to the daughter cells. A greater eccDNA copy number with an oncogene can be inherited [12]. Tumor cells engage in great adaptive mechanisms that help cancer cells survive in whatsoever conditions. For example, with respect to environmental conditions, the particular eccDNA number in tumors is changed, thereby helping tumor cells adapt with a different mechanism. This has been found to occur in glioblastomas, wherein mutation of EGFR is common, which results in the development of an oncogenic EGFRvIII variant. Tumor cell proliferation and growth are promoted by EGFRvIII, which also increases the sensitization of cancer cells to tyrosine kinase inhibitors [78]. Moreover, when cells lose DMs that carry mutated EGFR, they become resistant to the EGFR tyrosine kinase inhibitors [78]. It is also believed that extrachromosomal driver mutations are possible that occur when eccDNAs amplify, which proves vital for helping tumors evolve [81]. Thus, it can be said that the presence of eccDNA occurs more commonly in tumors resulting in adaption, heterogeneity of tumors, and thus their evolution (**Table 2**).

Apart from these functions, eccDNA has a role in aging, which is directly or indirectly linked to the tumors. In Yeast, the accumulation of eccDNAs containing ribosomal RNA genes occurs, thereby contributing to Yeast cell aging [88]. These ribosomal DNAs can replicate because of the presence of an Autonomously replicating sequence (ARS) sequence [88]. Furthermore, they show a tendency to transfer to mother cells during each cell division that leads to a higher accumulation of eccDNAs in aging mother cells. The lifespan of the daughter cell is prolonged with this lesser eccDNA concentration. However, researchers do not know the exact mechanism of how senescence is triggered by the eccDNAs, eventually leading to the mortality of aging cells. But it is presumed that it affects the replication and transcription mechanism by titrating different components involved [90]. The phenomenon that

| Properties | Chromosomal DNA | Extrachromosomal Circular DNA | References |
|---|---|---|---|
| Size | Much larger in size | Smaller | [24, 26, 37] |
| Stability | Highly stable | Not stable | [75] |
| Segregation | Segregates equally to daughter cells | Segregates unequally to daughter cells | [12] |
| Oncogene presence rate | Less number of oncogenes/ base pair | High number of oncogenes/ base pair | [12, 80] |
| Heterogeneity | Lesser | Higher | [80] |
| Oncogene amplification | Lesser | Higher | [12] |
| Effect on heterogeneity | Overcomes heterogeneity | Maintains heterogeneity | [80] |
| Oncogene survival | Lesser | Higher | [81] |
| Mutations | Tumour unfriendly | Tumour unfriendly | [81] |
| Occurrence | No significant difference between normal and cancer tissues | Occurs more commonly in cancer tissues | [12] |
| Harm | Oncogenes are less harmful, if present | Oncogenes are much harmful when present on eccDNA | [12] |
| Drug resistance | Lesser | Higher | [12] |
| Tumour cell diversity | Lesser | Higher | [12] |
| Tumour growth | Lower growth rate | Higher growth rate | [12] |

**Table 2.**
*Difference between chromosomal DNA and extrachromosomal circular DNA in promoting tumour survival and aggressiveness.*

eccDNAs occur abundantly in these cells may be responsible for senescence and mortality of aging cells. Considering this hypothesis, expressing ARS plasmid at an abnormal place is enough to arrest aging cells, which eventually leads to their death [90]. Since eccDNAs have been found to accumulate in normal cells, too, the discoveries have led scientists to scratch their heads whether accumulating levels of eccDNA cause aging in other higher eukaryotes or not. Besides relating to aging in Yeast, eccDNA has a similar role in mammals [90–92]. The concentrations of eccDNA in cells serve as an index of aging, as was found in senescence-resistant SAM-R mice, where they showed a higher amplification [93]. Another study supported this finding in which a stronger ERC load resulted due to sgs1 gene mutation, which associates with reduced life span and premature aging [90]. Another study studied the eccDNA replication in the cell cycle, where the concentrations increased in an exponential manner. This may be a sort of clock determining the yeast life span. In Yeast, the rise in the mortality rates also correlates with the increase in ERC numbers [90]. Some other studies focused on the relationship between the formation of eccDNA and transcriptional activity. It was found that certain genes that are sensitive to environmental stimuli are transcriptionally stimulated, which triggers protein-coding eccDNA generation in aging budding yeasts [94]. Similarly, under aging conditions, the structural characteristics of eccDNA were studied in cultured human lung fibroblasts and rat lymphocytes. The results showed a greater size of eccDNA, and high dispersion with the eccDNA almost doubles in number [91]. This means that the accumulation of eccDNA in tumors is primarily because of the upregulation of the synthesis rather than the downregulation of degradation.

In addition, other roles of eccDNA may also exist, such as gene compensation [95]. For instance, in Yeast two pairs of genes-HTA1-HTB1 and HTA2-HTB2 encode histones H2A and H2B, respectively. Upon deletion of HTA1-HTAB1, HTA2-HTB2 genes are amplified through dosage compensation by forming a naïve eccDNA. This eccDNA contains 39 kilobase pairs of the 2nd chromosome, including a centromere, HTA2-HTB2, H3-H4 locus of histones, and several origins of replication [95]. The creation of this naïve eccDNA occurs when two Ty1 retrotransposon elements recombine, flanking this portion [95]. Thus, the compensation for the H2A and H2B decrease occurs through elevation in HTA2-HTB2 eccDNA formation in these deleted strains of HTA1-HTB1. In comparison to DMs, the smaller types of eccDNAs occur more commonly, but we do not know much about them yet [46, 84]. Their size is little to have genes encoding proteins but larger enough to encode parts of genes or short RNAs with regulatory functions. Moreover, microDNAs have the property of acting as molecular sponges; indirect gene expression occurs when they sponge different transcription factors. More recently, it became clear that the microDNAs may occur in plasma and serum of both human beings and mice as circulating DNA [84]. In the technique or liquid biopsy, they may act as potential biomarkers. A novel cell communication mechanism may exist if microDNAs are present in other cells. This may be just a theoretical assumption right now but may open new fields of investigation in the future. Another important query to solve is how this eccDNA survives autoimmune pathways in the cytoplasm. If the naked DNA is present in the cell cytoplasm, then the cGAS pathway is activated, which stimulates the immune system through interferon expression [96, 97]. This is how our bodies respond to any foreign antigen entering our bodies and vital parts of the immune system. EccDNAs release may be stimulated during mitosis, which either leads to their degradation by TREX1 like enzymes or activation of cGAS pathways. Thus, it can be said that the eccDNAs are usual endogenous antigens if not having chromatin protection, which leads to activation of several autoimmune pathways.

### 3.2 Clinical utility

Tumors possess specific characteristics that help in their prognosis and identification. One of them is the presence of eccDNA, which may be an essential tool for tumor prognosis. However, having said that, normal tissues also release eccDNAs; thus, it becomes pivotal to identify these differences between normal and cancerous tissues. One of the ways to do so may be identifying the eccDNA length. For instance, research indicates that tumor cells of human origin show longer eccDNA molecules than those found in the normal mouse cell lines [46]. Similarly, both normal and cancerous lung tissues showed the presence of microDNA, with both the types showing the same known properties of eccDNA [84]. However, the eccDNA was removed and analyzed from human patients who have lung cancer, and the length of eccDNA was measured. It was found that eccDNA from the same patient shows variations in length, with tumor eccDNA larger in size than the normal cell eccDNA [84]. This property will certainly help researchers to differentiate cancerous and normal tissue and might prove a good biomarker. This will be true if eccDNA from normal and tumor cells show a predictable behavior. At the same time, many researchers have focused on liquid biopsy, which used high-throughput sequencing technology for the identification of linearized DNA fragments specific to tumors in the plasma or serum [78]. This discovery has rather been more recent [84]. Moreover, the microDNA obtained from mice, mice tissues, human tumors, and chicken showed the same characteristics as that of the microDNA obtained from circulation. These properties include distribution in genome, higher GC content, direct repeats that are 2–15 bases long and flank the source genomic sites, etc. Moreover, it is an intergenic and genic region that gives rise to eccDNA. As mentioned, the microDNA obtained from lung cancers was of greater length than microDNA from normal tissue of the same person. The surgery also affects the eccDNA length, as observed in some studies [84]. Researchers found longer circular DNA levels in patients prior to surgery and smaller lengths post-surgery (6 weeks after surgical resection) [84]. This indicates the stability of eccDNAs in comparison to linear DNA and thus may be advantageous to use eccDNA for the purpose of liquid biopsy.

## 4. Conclusion

The eccDNA has a great role in the evolution of cancer as driver oncogene amplification contributes to heterogeneity of tumors and resistance to drugs. What makes the condition worse is that driver oncogenes present on eccDNA prove to be more harmful than in chromosomal DNA. The frequency of eccDNA has been much more common than was earlier thought, helping tumors evolve and become resistant to various drugs. The latest research conducted on the topic also found that the presence of eccDNA in nearly all tumors is much more common, providing immunity against therapeutic drugs [12]. The team explored the eccDNA presence after analyzing cell lines from 17 different tumor types. This was done by taking metaphase chromosomes from 2572 dividing cells whose ECdetect (a software package for conducting unbiased analysis and detecting eccDNA) was developed to study structural and functional properties. They concluded that almost half of the human cancers showed eccDNA presence. Moreover, eccDNA aids in better driver oncogene amplification than the chromosomal DNA, thereby accelerating transcript level [12]. Importantly, it is clear that eccDNA has a much greater role in promoting drug resistance, diversity, and growth of the tumor cells than the same genes located on chromosomal loci. This explains how the evolution and diversification of cancers occur. Moreover, around 90% of the tumors (patient-derived) models show eccDNA presence, meaning that eccDNA is more likely to harbor cancer oncogenes than chromosomes. The copy number of an oncogene and intratumoural heterogeneity is increased more effectively by the eccDNA amplification as predicted by mathematical models [12]. Moreover, a growing number of studies are showing the role of eccDNAs in harboring proto-oncogenes. The eccDNA is not only offering them a suitable environment to sustain themselves but also making them resistant to numerous drugs. Different genes show a link with different cancers, such as DHFR gene with colon, cervical and breast cancer, CA125 gene with ovarian cancer, MDR1 gene with oral squamous cell carcinoma (OSCC), a c-Myc gene with colon cancer and leukemia, HER2 gene with breast cancer, MDM2 and EGFRvIII with glioblastoma, as depicted in **Table 3** [78, 98–106]. Thus, tumors maintain their high gene number and heterogeneity more easily with the help of eccDNA. Moreover, the distribution of eccDNA to the daughter cells occurs randomly, meaning that a tumor may have either whole or no eccDNA. This increases the variation in the copy number of oncogenes, which ultimately makes the tumor more heterogeneous in terms of

| Genes | Cancer type | References |
|---|---|---|
| DHFR | Colon cancer, cervical cancer, breast cancer | [98–101] |
| CA125 | Ovarian cancer | [102] |
| MDR1 | OSCC | [103] |
| c-Myc | Colon cancer, leukaemia | [104, 105] |
| HER2 | Breast cancer | [106] |
| MDM2 | Glioblastoma | [78] |
| EGFRvIII | Glioblastoma | [78] |

**Table 3.**
*Some common oncogenes that are contained on eccDNAs.*

resistance to any kind of environmental changes, like due to drugs. The more common discovery of eccDNA in different cancer types is surprising as researchers have been focusing more on which genes cause cancer rather than where these genes occur. Although some cancer biologists reported the eccDNA presence as early as the 1960s, the tools to quantify eccDNA were lacking. More studies are demanding to know the exact mechanisms of eccDNA formation and maintenance and how the Tumor Microenvironment changes eccDNA levels by altering its composition.

## Conflict of interest

The authors declare no conflict of interest.

## Other declarations/Notes

None.

## Abbreviations

| | |
|---|---|
| ARS | autonomously replicating sequence |
| DM | double minutes |
| dsDNA | double-stranded deoxyribonucleic acid |
| eccDNA | extrachromosomal circular DNA |
| EGFR | epidermal growth factor receptor |
| EIF5A2 | eukaryotic translation initiation factor 5A2 |
| FISH | fluorescence in situ hybridization |
| HER2 | human epidermal growth factor receptor 2 |
| MDM2 | mouse double minute 2 homolog |
| MDR1 | multidrug resistance 1 |
| OC | ovarian cancer |
| OSCC | oral squamous cell carcinoma |
| SB | Southern Blotting |
| ssDNA | single-stranded deoxyribonucleic acid |
| UTR | untranslated regions |
| WGS | whole-genome sequencing |

*DOI: http://dx.doi.org/10.5772/intechopen.102687*

## Author details

Parvaiz Yousuf
Department of Zoology, Central University of Kashmir, Ganderbal,
Jammu and Kashmir, India

*Address all correspondence to: saleemparvaiz444@gmail.com

## References

[1] Dahm R. Discovering DNA: Friedrich Miescher and the early years of nucleic acid research. Human Genetics. 2008;**122**(6):565-581. DOI: 10.1007/s00439-007-0433-0

[2] WATSON JD, CRICK FH. Genetical implications of the structure of deoxyribonucleic acid. Nature. 1953;**171**(4361):964-967. DOI: 10.1038/171964b0

[3] Saccone C, Gissi C, Lanave C, Larizza A, Pesole G, Reyes A. Evolution of the mitochondrial genetic system: An overview. Gene. 2000;**261**(1):153-159. DOI: 10.1016/s0378-1119(00)00484-4

[4] Sakamoto W, Takami T. Chloroplast DNA dynamics: Copy number, quality control and degradation. Plant & Cell Physiology. 2018;**59**(6):1120-1127. DOI: 10.1093/pcp/pcy084

[5] Hotta Y, Bassel A. Molecular size and circularity of dna in cells of mammals and higher plants. Proceedings of the National Academy of Sciences of the United States of America. 1965;**53**(2):356-362. DOI: 10.1073/pnas.53.2.356

[6] Martin DP, Biagini P, Lefeuvre P, Golden M, Roumagnac P, Varsani A. Recombination in eukaryotic single stranded DNA viruses. Viruses. 2011;**3**(9):1699-1738. DOI: 10.3390/v3091699 Epub 2011 Sep 13

[7] Hindmarsh P, Leis J. Retroviral DNA integration. Microbiology and Molecular Biology Reviews . 1999;63(4):836-843, table of contents. DOI: 10.1128/MMBR.63.4.836-843.1999

[8] Smith GL, Vaccinia virus, Brian WJ Mahy, Marc HV Van Regenmortel. Encyclopedia of Virology (3rd). USA Academic Press. 2008, 243-250. 9780123744104. 10.1016/B978-012374410-4.00525-2

[9] Drlica K, Gennaro ML, Plasmids. Sydney Brenner, Jefferey H. Miller, Encyclopedia of Genetics, Academic Press, USA 2001. 1485-1490. 9780122270802, 10.1006/rwgn.2001.1000

[10] Taanman JW. The mitochondrial genome: Structure, transcription, translation and replication. Biochimica et Biophysica Acta. 1999;**1410**:103-123

[11] Paulsen T, Shibata Y, Kumar P, Dillon L, Dutta A. Small extrachromosomal circular DNAs, microDNA, produce short regulatory RNAs that suppress gene expression independent of canonical promoters. Nucleic Acids Research. 2019;**47**:4586-4596

[12] Turner KM, Deshpande V, Beyter D, Koga T, Rusert J, Lee C, et al. Extrachromosomal oncogene amplification drives tumour evolution and genetic heterogeneity. Nature. 2017;**543**(7643):122-125. DOI: 10.1038/nature21356

[13] Storlazzi CT et al. Gene amplification as DMs or homogeneously staining regions in solid tumours: Origin and structure. Genome Research. 2010;**20**:1198-1206

[14] Tomaska L, Nosek J, Kramara J, Griffith JD. Telomeric circles: Universal players in telomere maintenance? Nature Structural & Molecular Biology. 2009;**16**:1010-1015

[15] Cohen S, Regev A, Lavi S. Small polydispersed circular DNA (spcDNA) in

human cells: Association with genomic instability. Oncogene. 1997;**14**:977-985

[16] Koche RP, Rodriguez-Fos E, Helmsauer K, Burkert M, MacArthur IC, Maag J, et al. Extrachromosomal circular DNA drives oncogenic genome remodeling in neuroblastoma. Nature Genetics. 2020;**52**(1):29-34. DOI: 10.1038/s41588-019-0547-z

[17] Stahl F. A chain model for chromosomes. Journal de Chimie Physique. 1961;**58**:1072-1077

[18] Møller HD, Parsons L, Jørgensen TS, Botstein D, Regenberg B. Extrachromosomal circular DNA is common in yeast. Proceedings of the National Academy of Sciences of the United States of America. 2015 Jun;**112**(24):E3114-E3122. DOI: 10.1073/pnas.1508825112

[19] Stanfield SW, Lengyel JA. Small circular DNA of Drosophila melanogaster: chromosomal homology and kinetic complexity. Proceedings of the National Academy of Sciences of the USA. 1979;**76**(12):6142-6146. doi: 10.1073/pnas.76.12.6142

[20] Sunnerhagen P, Sjöberg RM, Karlsson AL, Lundh L, Bjursell G. Molecular cloning and characterization of small polydisperse circular DNA from mouse 3T6 cells. Nucleic Acids Research. 1986;**14**(20):7823-7838. DOI: 10.1093/nar/14.20.7823

[21] Stanfield SW, Helinski DR. Cloning and characterization of small circular DNA from Chinese hamster ovary cells. Molecular and Cellular Biology. 1984;**4**(1):173-180. DOI: 10.1128/mcb.4.1.173-180

[22] Smith CA, Vinograd J. Small polydisperse circular DNA of HeLa cells. Journal of Molecular Biology. 1972 Aug;**69**(2):163-178. DOI: 10.1016/0022-2836(72)90222-7

[23] Motejlek K, Assum G, Krone W, Kleinschmidt AK. The size of small polydisperse circular DNA (spcDNA) in angiofibroma-derived cell cultures from patients with tuberous sclerosis (TSC) differs from that in fibroblasts. Human Genetics. 1991;**87**(1):6-10. DOI: 10.1007/BF01213083

[24] Motejlek K, Schindler D, Assum G, Krone W. Increased amount and contour length distribution of small polydisperse circular DNA (spcDNA) in Fanconi anemia. Mutation Research. 1993;**293**(3):205-214. DOI: 10.1016/0921-8777(93)90071-n

[25] Neidlinger C, Assum G, Krone W, Dietrich C, Hochsattel R, Klotz G. Increased amounts of small polydisperse circular DNA (spcDNA) in angiofibroma-derived cell cultures from patients with tuberous sclerosis (TS). Human Genetics. 1988;**79**(3):286-288. DOI: 10.1007/BF00366254

[26] Cox D, Yuncken C, AI S. Minute chromatin bodies in malignant tumours of childhood. Lancet. 1965;**1**(7402):55-58. DOI: 10.1016/s0140-6736(65)90131-5

[27] Spriggs AI, Boddington MM, Clarke CM. Chromosomes of human cancer cells. British Medical Journal. 1962;**2**(5317):1431-1435. DOI: 10.1136/bmj.2.5317.1431

[28] Radloff R, Bauer W, Vinograd J. A dye-buoyant-density method for the detection and isolation of closed circular duplex DNA: The closed circular DNA in HeLa cells. Proceedings of the National Academy of Sciences of the United States of America. 1967;**57**(5):1514-1521. DOI: 10.1073/pnas.57.5.1514

[29] Paulsen T, Kumar R, Koseoglu M.M, Dutta A, Discoveries

of extrachromosomal circles of DNA in Normal and tumour cells, Trends in Genetics. 2018;**34**(4);270-278

[30] Cohen S, Lavi S. Induction of circles of heterogeneous sizes in carcinogen-treated cells: Two-dimensional gel analysis of circular DNA molecules. Molecular and Cellular Biology. 1996;**16**(5):2002-2014. DOI: 10.1128/MCB.16.5.2002

[31] Agsteribbe E, Kroon AM, van Bruggen EF. Circular DNA from mitochondria of neurospora crassa. Biochimica et Biophysica Acta. 1972;**269**(2):299-303. DOI: 10.1016/0005-2787(72)90439-x

[32] Billheimer FE, Avers CJ. Nuclear and mitochondrial DNA from wild-type and petite yeast: Circularity, length, and buoyant density. Proceedings of the National Academy of Sciences of the United States of America. 1969;**64**(2):739-746. DOI: 10.1073/pnas.64.2.739

[33] Buongiorno-Nardelli M, Amaldi F, Lava-Sanchez PA. Electron microscope analysis of amplifying ribosomal DNA from xenopus laevis. Experimental Cell Research. 1976;**98**(1):95-103. DOI: 10.1016/0014-4827(76)90467-5

[34] Ono T, Ozeki Y, Okubo S, Inoki S. Characterization of nuclear and satellite DNA from trypanosomes. Biken Journal. 1971;**14**(3):203-215

[35] Stanfield S, Helinski DR. Small circular DNA in Drosophila melanogaster. Cell. Oct 1976;**9**(2):333-345. DOI: 10.1016/0092-8674(76)90123-9

[36] Wong FY, Wildman SG. Simple procedure for isolation of satellite DNAs from tobacco leaves in high yield and demonstration of minicircles. Biochimica et Biophysica Acta. 1972;**259**(1):5-12. DOI: 10.1016/0005-2787(72)90468-6

[37] Bertelsen AH, Humayun MZ, Karfopoulos SG, Rush MG. Molecular characterization of small polydisperse circular deoxyribonucleic acid from an African green monkey cell line. Biochemistry. 1982;**21**(9):2076-2085. DOI: 10.1021/bi00538a015

[38] Van Loon N, Miller D, Murnane JP. Formation of extrachromosomal circular DNA in HeLa cells by nonhomologous recombination. Nucleic Acids Research. 1994;**22**(13):2447-2452. DOI: 10.1093/nar/22.13.2447

[39] Gaubatz JW, Flores SC. Purification of eucaryotic extrachromosomal circular DNAs using exonuclease III. Analytical Biochemistry. 1 Feb 1990;**184**(2):305-310. DOI: 10.1016/0003-2697(90)90685-3

[40] Cohen S, Mechali M. A novel cell-free system reveals a mechanism of circular DNA formation from tandem repeats. Nucleic Acids Research. 2001;**29**(12):2542-2548. DOI: 10.1093/nar/29.12.2542

[41] Cohen S, Agmon N, Sobol O, Segal D. Extrachromosomal circles of satellite repeats and 5S ribosomal DNA in human cells. Mobile DNA. 2010;**1**(1):11. DOI: 10.1186/1759-8753-1-11

[42] Navrátilová A, Koblízková A, Macas J. Survey of extrachromosomal circular DNA derived from plant satellite repeats. BMC Plant Biology. 2008;**8**:90. DOI: 10.1186/1471-2229-8-90

[43] Beland JL, Longo JA, Hahn PJ. CpG island mapping of a mouse double-minute chromosome. Molecular and Cellular Biology. 1993;**13**(8):4459-4464. DOI: 10.1128/mcb.13.8.4459-4464.1993

[44] Carroll SM, Gaudray P, De Rose ML, Emery JF, Meinkoth JL, Nakkim E, et al. Characterization of an episome produced in hamster cells that amplify a

transfected CAD gene at high frequency: Functional evidence for a mammalian replication origin. Molecular and Cellular Biology. 1987;**7**(5):1740-1750. DOI: 10.1128/mcb.7.5.1740-1750.1987

[45] Ståhl F, Wettergren Y, Levan G. Amplicon structure in multidrug-resistant murine cells: A nonrearranged region of genomic DNA corresponding to large circular DNA. Molecular and Cellular Biology. 1992;**12**(3):1179-1187. DOI: 10.1128/mcb.12.3.1179-1187.1992

[46] Shibata Y, Kumar P, Layer R, Willcox S, Gagan JR, Griffith JD, et al. Extrachromosomal microDNAs and chromosomal microdeletions in normal tissues. Science. 2012;**336**(6077):82-86. DOI: 10.1126/science.1213307 Epub 2012 Mar 8. Erratum in: Science. 2012 Jun 22;336(6088):1506

[47] Dillon LW, Kumar P, Shibata Y, Wang YH, Willcox S, Griffith JD, et al. Production of extrachromosomal MicroDNAs is linked to mismatch repair pathways and transcriptional activity. Cell Reports. 2015;**11**(11):1749-1759. DOI: 10.1016/j.celrep.2015.05.020

[48] Cohen S, Yacobi K, Segal D. Extrachromosomal circular DNA of tandemly repeated genomic sequences in drosophila. Genome Research. 2003;**13**:1133-1145

[49] Cohen S, Houben A, Segal D. Extrachromosomal circular DNA derived from tandemly repeated genomic sequences in plants. The Plant Journal. 2008;**53**:1027-1034

[50] Zhu J, Yu Y, Meng X, Fan Y, Zhang Y, Zhou C, et al. De novo-generated small palindromes are characteristic of amplicon boundary junction of double minutes. International Journal of Cancer. 2013;**133**:797-806

[51] Barr FG, Nauta LE, Davis RJ, Schafer BW, Nycum LM, Biegel JA. In vivo amplification of the PAX3-FKHR and PAX7-FKHR fusion genes in alveolar rhabdomyosarcoma. Human Molecular Genetics. 1996;**5**:15-21

[52] Rodley P, McDonald M, Price B, Fright R, Morris C. Comparative genomic hybridization reveals previously undescribed amplifications and deletions in the chronic myeloid leukemia-derived K-562 cell line. Genes, Chromosomes & Cancer. 1997;**19**:36-42

[53] Van Roy N, Vandesompele J, Menten B, Nilsson H, De Smet E, Rocchi M, et al. Translocation-excision-deletion-amplification mechanism leading to nonsyntenic coamplification of MYC and ATBF1. Genes, Chromosomes & Cancer. 2006;**45**:107-117

[54] Korbel JO, Campbell PJ. Criteria for inference of chromothripsis in cancer genomes. Cell. 2013;**152**:1226-1236

[55] Zhang CZ, Spektor A, Cornils H, Francis JM, Jackson EK, Liu S, et al. Chromothripsis from DNA damage in micronuclei. Nature. 2015;**522**:179-184

[56] Mc CB. Chromosome organization and genic expression. Cold Spring Harbor Symposia on Quantitative Biology. 1951;**16**:13-47

[57] Vukovic B, Beheshti B, Park P, Lim G, Bayani J, Zielenska M, et al. Correlating breakage-fusion-bridge events with the overall chromosomal instability and in vitro karyotype evolution in prostate cancer. Cytogenetic and Genome Research. 2007;**116**:1-11

[58] Zakov S, Kinsella M, Bafna V. An algorithmic approach for breakage-fusionbridge detection in tumor genomes. Proceedings of the National Academy of Sciences of the United States of America. 2013;**110**:5546-5551

[59] Murnane JP, Sabatier L. Chromosome rearrangements resulting from telomere dysfunction and their role in cancer. BioEssays. 2004;**26**:1164-1174

[60] Carroll SM, DeRose ML, Gaudray P, Moore CM, Needham-Vandevanter DR, Von Hoff DD, et al. Double minute chromosomes can be produced from precursors derived from a chromosomal deletion. Molecular and Cellular Biology. 1988;**8**(4):1525-1533. DOI: 10.1128/mcb.8.4.1525-1533.1988

[61] Moller HD, Mohiyuddin M, Prada-Luengo I, Sailani MR, Halling JF, Plomgaard P, et al. Circular DNA elements of chromosomal origin are common in healthy human somatic tissue. Nature Communications. 2018;**9**:1069

[62] Moller HD, Parsons L, Jorgensen TS, Botstein D, Regenberg B. Extrachromosomal circular DNA is common in yeast. Proceedings of the National Academy of Sciences of the United States of America. 2015;**112**:E3114-E3122

[63] Bao Y, Liu J, You J, Wu D, Yu Y, Liu C, et al. Met promotes the formation of double minute chromosomes induced by Sei-1 in NIH-3T3 murine fibroblasts. Oncotarget. 2016;**7**:56664-56675

[64] Guan XY, Fung JM, Ma NF, Lau SH, Tai LS, Xie D, et al. Oncogenic role of eIF-5A2 in the development of ovarian cancer. Cancer Research. 2004;**64**:4197-4200

[65] Shoura MJ, Gabdank I, Hansen L, Merker J, Gotlib J, Levene SD, et al. Intricate and Cell Type-Specific Populations of Endogenous Circular DNA (eccDNA) in Caenorhabditis elegans and Homo sapiens. G3 (Bethesda). 2017;**7**(10):3295-3303. DOI: 10.1534/g3.117.300141

[66] Vogt N, Gibaud A, Lemoine F, de la Grange P, Debatisse M, Malfoy B. Amplicon rearrangements during the extrachromosomal and intrachromosomal amplification process in a glioma. Nucleic Acids Research. 2014;**42**(21):13194-13205. DOI: 10.1093/nar/gku1101

[67] deCarvalho AC, Kim H, Poisson LM, et al. Discordant inheritance of chromosomal and extrachromosomal DNA elements contributes to dynamic disease evolution in glioblastoma. Nat Genet. 2018;**50**:708-717. DOI: 10.1038/s41588-018-0105-0

[68] Nowell PC. The clonal evolution of tumour cell populations. Science. 1976;**194**(4260):23-28. DOI: 10.1126/science.959840

[69] McGranahan N, Swanton C. Biological and therapeutic impact of intratumour heterogeneity in cancer evolution. Cancer Cell. 2015;**27**:15-26

[70] Marusyk A, Almendro V, Polyak K. Intra-tumour heterogeneity: A looking glass for cancer? Nature Reviews. Cancer. 2012;**12**:323-334

[71] Yates LR, Campbell PJ. Evolution of the cancer genome. Nature Reviews. Genetics. 2012;**13**:795-806

[72] Greaves M, Maley CC. Clonal evolution in cancer. Nature. 2012;**481**:306-313

[73] Von Hoff DD, Needham-VanDevanter DR, Yucel J, Windle BE, Wahl GM. Amplified human MYC oncogenes localized to replicating submicroscopic circular DNA molecules. Proceedings of the National Academy of Sciences of the United States of America. 1988;**85**:4804-4808

[74] Garsed DW, Marshall OJ, Corbin VD, Hsu A, Di Stefano L, Schröder J,

et al. The architecture and evolution of cancer neochromosomes. Cancer Cell. 2014;**26**(5):653-667. DOI: 10.1016/j.ccell.2014.09.010

[75] Mitelman F, Johansson B, Mertens F. Mitelman Database of Chromosome Aberrations and Gene Fusions in Cancer. The Lund University; 2016. Available from: https://www.clinicalgenetics.lu.se/division-clinical-genetics/database-chromosome-aberrations-and-gene-fusions-cancer

[76] Sanborn JZ et al. Double minute chromosomes in glioblastoma multiforme are revealed by precise reconstruction of oncogenic amplicons. Cancer Research. 2013;**73**:6036-6045

[77] Zack TI et al. Pan-cancer patterns of somatic copy number alteration. Nature Genetics. 2013;**45**:1134-1140

[78] Nathanson DA, Gini B, Mottahedeh J, Visnyei K, Koga T, Gomez G, et al. Targeted therapy resistance mediated by dynamic regulation of extrachromosomal mutant EGFR DNA. Science. 2014 Jan 3;**343**(6166):72-76. DOI: 10.1126/science.1241328

[79] Windle B, Draper BW, Yin YX, O'Gorman S, Wahl GM. A central role for chromosome breakage in gene amplification, deletion formation, and amplicon integration. Genes & Development. 1991;**5**:160-174

[80] Andor N et al. Pan-cancer analysis of the extent and consequences of intratumour heterogeneity. Nature Medicine. 2016;**22**:105-113

[81] Li X et al. Temporal and spatial evolution of somatic chromosomal alterations: A case-cohort study of Barrett's esophagus. Cancer Prevention Research (Philadelphia, Pa.). 2014;**7**:114-127

[82] Mishra S, Whetstine JR. Different facets of copy number changes: Permanent, transient, and adaptive. Molecular and Cellular Biology. 2016;**36**:1050-1063

[83] Schimke RT, Kaufman RJ, Alt FW, Kellems RF. Gene amplification and drug resistance in cultured murine cells. Science. 1978;**202**:1051-1055

[84] Nikolaev S et al. Extrachromosomal driver mutations in glioblastoma and low-grade glioma. Nature Communications. 2014;**5**:5690

[85] Biedler JL, Schrecker AW, Hutchison DJ. Selection of chromosomal variant in amethopterinresistant sublines of leukemia L1210 with increased levels of dihydrofolate reductase. Journal of the National Cancer Institute. 1963;**31**:575-601

[86] Kumar P, Dillon LW, Shibata Y, Jazaeri AA, Jones DR, Dutta A. Normal and cancerous tissues release extrachromosomal circular DNA (eccDNA) into the circulation. Molecular Cancer Research: MCR. 2017;**15**(9):1197-1205

[87] Vogt N et al. Molecular structure of double-minute chromosomes bearing amplified copies of the epidermal growth factor receptor gene in gliomas. Proceedings of the National Academy of Sciences of the United States of America. 2004;**101**(31):11368-11373

[88] Zuberi L et al. Rapid response to induction in a case of acute promyelocytic leukemia with MYC amplification on double minutes at diagnosis. Cancer Genetics and Cytogenetics. 2010;**198**(2):170-172

[89] Del Rey J, Prat E, Ponsa I, Lloreta J, Gelabert A, Algaba F, et al. Centrosome clustering and cyclin D1 gene amplification in double minutes are common events in chromosomal unstable bladder tumours. BMC Cancer. 2010;**10**:280. DOI: 10.1186/1471-2407-10-280

[90] Sinclair DA, Guarente L. Extrachromosomal rDNA circles--a cause of aging in yeast. Cell. 1997;**91**(7):1033-1042

[91] Kunisada T, Yamagishi H, Ogita Z, Kirakawa T, Mitsui Y. Appearance of extrachromosomal circular DNAs during in vivo and in vitro ageing of mammalian cells. Mechanisms of Ageing and Development. 1985;**29**:89-99

[92] Gaubatz JW, Flores SC. Tissue-specific and age-related variations in repetitive sequences of mouse extrachromosomal circular DNAs. Mutation Research. 1990;**237**:29-36

[93] Yamagishi H, Kunisada T, Takeda T. Amplification of extrachromosomal small circular DNAs in a murine model of accelerated senescence. A brief note. Mechanisms of Ageing and Development. 1985;**29**:101-103

[94] Hull RM, King M, Pizza G, Krueger F, Vergara X, Houseley J. Transcription-induced formation of extrachromosomal DNA during yeast ageing. PLoS Biology. 2019;**17**:e3000471

[95] Libuda DE, Winston F. Amplification of histone genes by circular chromosome formation in Saccharomyces cerevisiae. Nature. 2006;**443**(7114):1003-1007

[96] Mackenzie KJ, Carroll P, Martin CA, Murina O, Fluteau A, Simpson DJ, et al. cGAS surveillance of micronuclei links genome instability to innate immunity. Nature. 2017 Aug 24;**548**(7668):461-465. DOI: 10.1038/nature23449

[97] de Oliveira Mann CC, Kranzusch PJ. cGAS conducts micronuclei DNA surveillance. Trends in Cell Biology. 2017

[98] Meng X, Qi X, Guo H, Cai M, Li C, Zhu J, et al. Novel role for nonhomologous end joining in the formation of double minutes in methotrexate-resistant colon cancer cells. Journal of Medical Genetics. 2015;**52**(2):135-144

[99] Cai M, Zhang H, Hou L, Gao W, Song Y, Cui X, et al. Inhibiting homologous recombination decreases extrachromosomal amplification but has no effect on intrachromosomal amplification in methotrexate-resistant colon cancer cells. International Journal of Cancer. 2019;**144**(5):1037-1048

[100] Ruiz-Herrera A, Smirnova A, Khoriauli L, Nergadze SG, Mondello C, Giulotto E. Gene amplification in human cells knocked down for RAD54. Genome Integrity. 2011;**2**(1):5

[101] Hahn P, Nevaldine B, Morgan WF. X-ray induction of methotrexate resistance due to dhfr gene amplification. Somatic Cell and Molecular Genetics. 1990;**16**(5):413-423

[102] Zhang CY, Feng YX, Yu Y, Sun WJ, Bai J, Chen F, et al. The molecular mechanism of resistance to methotrexate in mouse methotrexate-resistant cells by cancer drug resistance and metabolism SuperArray. Basic & Clinical Pharmacology & Toxicology. 2006;**99**(2):141-145

[103] Von Hoff DD, Waddelow T, Forseth B, Davidson K, Scott J, Wahl G. Hydroxyurea accelerates loss of extrachromosomally amplified genes from tumor cells. Cancer Research. 1991;**51**(23 Pt 1):6273-6279

[104] Shimizu N, Itoh N, Utiyama H, Wahl GM. Selective entrapment of

extrachromosomally amplified DNA by nuclear budding and micronucleation during S phase. The Journal of Cell Biology. 1998;**140**(6):1307-1320

[105] Eckhardt SG, Dai A, Davidson KK, Forseth BJ, Wahl GM, Von Hoff DD. Induction of differentiation in HL60 cells by the reduction of extrachromosomally amplified c-myc. Proceedings of the National Academy of Sciences of the United States of America. 1994;**91**(14):6674-6678

[106] Vicario R, Peg V, Morancho B, Zacarias-Fluck M, Zhang J, MartinezBarriocanal A, et al. Patterns of HER2 gene amplification and response to anti-HER2 therapies. PLoS One. 2015;**10**(6):e0129876

Chapter 8

# Genetics of Colorectal Cancer Racial Disparities

*Jennie Williams, Jenny Paredes and Shrey Thaker*

## Abstract

This chapter describes genetics and epigenetics discoveries that have allowed investigators to better define cancer at the molecular level. Taking into consideration the expanse of the field of cancer, the focus will be on colon cancer as a platform to provide examples of techniques, recent discoveries, and translation of genetic studies to cancer care. In addition, this segment contributes to our understanding of racial and ethnic disparities in colon cancer and the use of -omic assessments as an application in cancer research. Thus, this section will provide an overarching view of cancer by defining the molecular characteristics of colon cancer; parameters of cancer disparities; and genetic factors that contribute to colon-tumor biology, specifically recent findings at the DNA, RNA, and protein levels. Importantly, the correlation of these factors with the immune system will be defined. This section ends with future directions for studying colon cancer in patients from medically underserved communities. In summary, this unit provides an introduction to how genetic and genomic investigations are helping to elucidate biological questions in an inclusive manner that will benefit patients on a global scale.

**Keywords:** colon cancer, disparities, African Americans, genetics, genomics, immunology

## 1. Introduction

According to the National Cancer Institute of The United States, cancer is defined as a disease in which cells grow uncontrollably and spread to other parts of the body [1]. Cancer can occur in any human tissue when cells lose control of cell division and multiply abnormally. Tumors, the accumulation of abnormal cells, can be limited to one location or invade into nearby tissues to form new tumors, a process known as metastasis. Cancerous tumors can be solid tumors, like colon cancer, or cancers of the blood, such as leukemias [1]. Regardless of the classification or tumor type, however; all cancers have common molecular features that have been identified as the hallmarks of cancer. These hallmarks encompass the biological abnormalities that are present in a cell to be classified as a cancer cell. They include the morphology, biology, metabolism, and genetic composition that are shared by all tumors.

The system used to organize the complexity of a cancer cell into simple hallmarks has been evolving for decades. Currently, there are 10 hallmarks of cancer. This includes self-sufficiency in growth signals, insensitivity to anti-growth signals, evading apoptosis, limitless replicative potential, sustained angiogenesis, tissue invasion, and metastasis, reprogramming energy metabolism and evading immune response, genome instability and mutation, and tumor-promoting inflammation [2]. Collectively, these characteristics arise from not only studying the biology but also assessing the genetic and genomic processes of cancer cells. For example, many hallmarks are common to both benign (non-cancerous tumors) and malignant growths, such as the evasion of apoptosis or cell death and limitless replication potential. Thus, it is only through genetic and genomic studies that the identification of additional markers provides evidence that cancer cells share the presence of genetic mutations and genomic instability. This approach is called the mutation theory and it argues that carcinogenesis is a process that initiates with genetic mutations that allow for the hallmarks of cancer to develop in a cell lifespan [3]. Therefore, we will summarize genetic factors that contribute to the hallmarks of cancer.

First, we will address the selective growth and proliferative advantage of cancer cells. Normal cells depend on growth signaling of a strictly regulated cell cycle to proliferate and maintain tissue homeostasis. On the other hand, in cancer cells, the growth and proliferative signals are altered by mutations in genes that code for growth ligands, receptors, and other survival-signaling molecules involved in apoptosis [4]. Depending on the biological role that they play in proliferation, growth factors can be upregulated or downregulated. Increased levels may enhance tumor progression, whereas, lower than normal levels may result in the escape of the tumor from regulation. A well-studied example is that of the transforming growth factor-β (TGF-β), which can be an anti-growth ligand but has also been implicated in tumor progression through stimulating differentiation of cancer cells. The duality of genes like TGF-β can be the result of gene amplification, somatic mutations, or chromosomal translocations that may lead to fusion proteins and aberrant signaling [4]. A more complex example is the rat sarcoma virus (RAS) protein, which is active in 30% of all cancers. This protein is often altered as the result of missense mutations in its gene or inactivating mutations in one of its negative regulators, resulting in a variety of effects such as enhanced growth and proliferation, suppression of apoptosis, rewiring of metabolism, promoting angiogenesis, and immune evasion. Thus, mutations in the *RAS* genetic pathway can be implicated in multiple hallmarks of cancer [5].

Similarly, another key regulator of cell growth is the tumor protein 53 (TP53), which is the most often mutated cancer gene, altered in over 50% of sequenced tumors. The main function of *TP53* is to detect cellular abnormalities that include genotoxic stress, excessive signaling, nutrient deprivation, and hypoxia. In response to these triggers, *TP53* can stop cell proliferation, initiate DNA repair mechanisms, or activate terminal differentiation and apoptosis. Not surprisingly, the genetic alterations to this tumor suppressor gene are involved in virtually all hallmarks of cancer [6]. Therefore, mutations and genetic alterations are important contributors to abnormal cell signaling, proliferation, and inhibition of cell death-regulators.

In addition to unregulated growth and cell death, cancer cells can develop the potential to invade and proliferate outside their original tumor niche. Metastasis is the process that allows cancer cells to form secondary tumors and metastatic disease is responsible for over 90% of cancer-related deaths and involves several steps. For a cancer cell to become metastatic, it must invade through the extracellular matrix, promote angiogenesis and tumor vasculature, survive transport in circulation, and manipulate foreign microenvironments [7]. Most human carcinoma cells migrate collectively in an aberrant pattern. In the case of solid tumors, such as those seen in colon cancer, cancer evolves from epithelial cells that are normally immotile and tightly adherent to one another and the surrounding matrix. These cells acquire mobility by an epithelial-mesenchymal transition (EMT) that allows an epithelial cell to become mesenchymal [8]. The biological changes of EMT can include mutations on genes that are involved in epithelial growth factors, tissue hypoxia, metabolic and mechanical stress, and matrix composition. Mutations in EMT transcription factors could result in repressing epithelial genes and activating mesenchymal ones, or in epigenetic modifications that facilitate cancer cell invasion and migration. Once cells acquire the ability to invade new tissues, they adapt by proliferating in their new microenvironment.

One of the main strategies for cancer cells to thrive in new microenvironments and to promote cancer growth in their primary niche is to accumulate genetic and epigenetic modifications that are advantageous for metabolic rewiring. Metabolism in cancer cells may include changes in the use of glucose, amino acids, nitrogen, and alterations in metabolic gene regulation [9]. Among many nutrients, the most relevant ones are glucose and glutamine as they play a crucial role in carbon degradation, synthesis of macromolecules, ATP generation, nitrogen uptake, and nucleotide biosynthesis [10]. In the case of glucose, cancer cells have developed the "Warburg effect;" the increased utilization of glucose under aerobic conditions. This effect results from genetic and epigenetic changes that increase the transport and degradation of glucose, as well as in the deregulation of signaling pathways such as PI3K/Akt and the oncogenes KRAS proto-oncogene GTPase (*KRAS*) and the proto-oncogene serine/threonine kinase (*BRAF*). Although less studied, the increased demand of glutamine by cancer cells appears to be involved in the processes of protein synthesis, protein degradation under nutrient-deprivation conditions, engulfment and digestion of living cells, and phagocytosis of apoptotic products [11]. Cancer cells rely not only on genetic and epigenetic changes to promote metastasis but also on their interactions with the neighboring cells. Stromal cells such as fibroblasts contribute to the intratumoral cell heterogeneity, where cancer and normal cells will contribute to tumor growth and progression. Cancer-associated fibroblasts contribute to therapeutic resistance, the acquisition of nutrients, and evasion of the immune system [12].

Finally, a hallmark of all cancer types is the ability to evade immune surveillance. The immune system uses cancer-immunoediting to regulate and eliminate cells that proliferate uncontrollably. Immunoediting is made up of three phases: elimination, equilibrium, and escape. Elimination of cancer cells is the ultimate effect of the immune surveillance of the innate and adaptive immune systems. There must be a recognition of tumor cells by innate immune cells, a maturation and migration of antigen-presenting cells, the generation of tumor-antigen-specific T-lymphocytes, the activation of cytotoxic mechanisms, and, finally, elimination of tumor cells [13]. For cancer immunoediting to be efficient, the immune system should be able to generate the genetic and epigenetic changes that are required to interact with tumor cells that undergo antigen remodeling and selection. Cancer cells with reduced immunogenicity result in the production of resistant variants with mutations that increase resistance to immune cytotoxicity. These tumor variants are characterized by genetic and epigenetic alterations that reduce tumor-antigen recognition, increase resistance to cell death, and induce immunological tolerance [14]. Furthermore, and in correlation with other hallmarks of cancer, cancer cells can suppress the cytotoxic components of the immune system through the secretion of immunosuppressive factors and inflammation. For example, cancer and stromal cells can secrete proinflammatory molecules, tumor-derived exosomes can suppress the function of immune cells, and

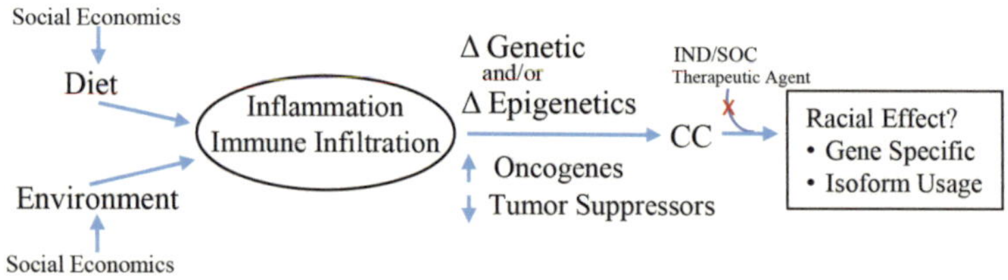


**Figure 1.**
*Differences between AA and CA colon cancer patients at various levels of genomic expression and control; specifically, gene expression, DNA methylation, and chemotherapeutic response. Upstream factors are associated with and driven by social determinates of health. IND (investigational new drugs), SOC (standard of care), and NA (normal adjacent).*

metabolic rewiring and altered microbiome can result in negative regulation of the immune system [14]. In summary, the immune modulation observed in tumors is currently recognized as a key player during cancer initiation and progression, and as a promising field for therapeutic manipulation.

Importantly, a major disparity in the prevalence, incidence and mortality of colorectal cancer (CRC) between African Americans (AA) and Caucasian Americans (CA) exists. Differences in response to treatment is an established factor influencing the overall survival of CRC patients. Here, we will address tumor biology and the importance of inclusivity in research. Knowledge of the molecular differences in CRC arising in different populations is needed to drive new therapeutic strategies and help overcome treatment resistance mechanisms to reduce disparities observed for this disease in minority populations. Depicted in **Figure 1** is an overview of those factors associated with cancer initiation, progression, and health disparity. Although we focus in this report on tumor biology, it is important to note that social determinants of health factor strongly into health disparity. Therefore, it is imperative that we not only understand tumor biology but are also able to address social determinates of health with respect to overall survival of all colon cancer patients.

## 2. Types of cancer

As it was mentioned in the introduction, cancer can arise from any cell in the human body that grows in an uncontrollable manner [1]. When classified by the location, or tissue, of origin, there about 200 different kinds of cancers. Nevertheless, when we focus on the aberrant gene expression of cancer and how it relates to the cellular composition, it can be concluded that there are 5 types of cancer: carcinoma, sarcoma, leukemia, lymphoma and myeloma, and brain and spinal cord [1].

Carcinomas can develop from the epithelial cells from the skin or the tissue lining of internal organs. Therefore, they could be further classified depending on the "skin layer" where they are localized, namely as adenocarcinomas, basal cell carcinomas, squamous carcinomas and transitional cell carcinomas. Another type of cancer—sarcoma—arises from connective tissue such as bone, cartilage, adipose tissue, muscles and vascular tissue (i.e., blood vessels). Leukemia, on the other hand, is characterized by developing specifically in white blood cells and its primary source, the bone marrow. The lymphomas and myelomas, develop from cells from the immune system, from locations such as the lymph nodes and the spleen. Lastly, cancers from the central nervous system, will be present in the brain and the spinal cord [1].

Carcinomas are the most common type of cancer, with 85% of the cancer cases in the United States, and include lung and colon cancer, which are number one and third in terms of incidence, respectively [1]. Sarcomas on the other hand, compose less than 1% of the cancer cases and most of the diagnosis are bone sarcomas [1]. Although leukemias are only the 3% of cancer cases in the United States, they are the most common type of cancer in children [1]. Cancers of the lymphatic system—lymphomas and myelomas—are 5% and 3% of the cancer cases, respectively. These cancers are particularly challenging to treat as they arise from abnormalities in the bone marrow, and they commonly require bone marrow transplantation [1]. Finally, brain and spinal tumors are 3% of the cancer cases in the USA and the most frequent subtype is the brain tumor from glial cells [1]. Taken together, it can be concluded that there are several cancer types and subtypes, based on the cellular composition of the cancer source. It is important to be noted, however, that many cancer types share aberrant gene expression that can cross over cancer types and subtypes. Therefore, we will address the shared genetic transgressions across the most common cancer types.

## 3. Aberrant gene expression in cancer

To address the observed abnormalities of gene expression across cancer types, we will start by describing the molecular basis for cancer progression. Mutations that lead to oncogene (stimulators of cell division) activation and tumor suppressor (regulators of cell cycle) dysregulation influence cancer initiation and progression. Such aberrant gene expression will, in turn, contribute to the hallmarks of cancer: unregulated growth and cell death. It is important to emphasize that the functional distinctions of mutations on the same gene from one cancer classification to another relies on the cellular differences and the distinct pathways that are deregulated in each cancer type.

Not surprisingly, most oncogenes and tumor suppressor genes are components of pathways that are involved in cell signaling. They are responsible for the regulation and generation of molecular signals, such as proteins, receptors, ligands, etc. For example, mutations in receptors of the tyrosine kinases RAS family, may act as oncogenes that are present in up to 80% of carcinomas [4]. Other signaling molecules that normally act as tumor suppressors, such as the TGFβ family, can lose regulatory function and commonly present as constitutively active receptors due to mutation [4]. An important example of aberrant gene expression in key pathways of cellular proliferation are those that are involved in DNA repair and cell division. A clear point of reference is the tumor suppressor gene *RB*. Mutations in the *RB* gene result in defects in the RB protein that normally acts as a restrictive molecule for cells to enter the S phase in the cell division cycle, leading to cancer cells proliferating inappropriately [4]. Interestingly, aberrant gene expression of the *RB* gene of the RB-regulatory pathway, including cyclins and cyclin-dependent kinases, are shared by many cancer types. The alterations of this pathway can be found in drastically different types of cancer: brain cancers (i.e., glioblastomas and carcinomas) and breast cancer, for example. Such observations highlight the importance of understanding gene pathways and their mechanism of action at the molecular and cellular level for improved targeted clinical outcomes.

The extent of the contribution of a single gene to the development of different cancer types is best exemplified by the aberrant mutations in the *TP53* gene [6]. As it was mentioned in the introduction, mutations in the *TP53* gene allow cells to survive and proliferate despite DNA damage. This tumor suppressor gene is present in 85% of human cancers and it is arguably the most important gene in human cancer, regardless of the cancer type. Hence, aberrant gene expression of *TP53* can simultaneously dysregulate the cell cycle, apoptotic signaling, and the overall genetic stability of a cell. The cascade reactions that are produced by abnormalities in the *TP53* pathway include alterations in the p21 pathway, the CDK complex, the MMR DNA repair system, among many others, making this gene relevant in virtually any cancer type [6].

Taken together the hallmarks of cancer and overlapping aberrant alterations in gene expression across cancer types, it is appropriate to conclude that using a type of cancer as a model of study, could allow us to better understand the genetics of cancer as a whole. Thus, we have selected colon cancer to illustrate some of the general principles and molecular mechanisms of tumor progression due to aberrant gene expression. Considering the global prevalence of colon cancer, responsible of about 11% of the cancer deaths, along with its defined stepwise genetic hallmark timeline, this cancer type seems to be the perfect prototype to address cancer gene expression. In addition, colon cancer provides researchers with the opportunity to study the physical progression of a tumor in the human epithelium along with the molecular changes that result from aberrations at the gene expression level.

## 4. Genetic factors that contribute to colon cancer

Colorectal cancer (CRC) is one of the most common cancers across the globe. It is estimated that the incidence of colon cancer will increase by 60% in several countries, positioning CRC as the second deadliest cancer type [15]. Up to 90% of CRC tumors arise from adenocarcinomas from the colon and rectum and up to 65% of the cases are sporadic (without a family history of CRC) [16]. Thus, the majority of CRC tumors progress from somatic and epigenetic alterations from modifiable risk factors such as metabolic comorbidities (e.g., obesity), diet, smoking, and alcohol consumption, among others [16]. It is important to highlight, however, that regardless of their inherited or somatic nature, several genetic factors and pathways have been identified in the pathogenesis of CRC.

In cases of hereditary CRC tumors, approximately 5% are classified as familial adenomatous polyposis (FAP) induced by alterations in the adenomatous polyposis coli (*APC*), MutL homolog 1 (*MLH1*), or the MutS homolog (*MSH2*) genes [17]. These inherited genetic factors are the result of chromosomal instability (loss or gain of chromosomal segments), aberrant methylation (altered CpG islands), and (or) microsatellite instability (MSI) due to the loss of DNA repair machinery [18]. These genetic alterations can also arise from somatic mutations (non-hereditary) of tumor-associated genes, for instance cell cycle regulators, which will in turn contribute to the aforementioned hallmarks, tumor initiation and progression [18].

Hence, either from hereditary factors or from genetic changes during the life span of the patient, CRC is determined by several genetic pathways. Although CRC tumors are considered heterogeneous at the molecular level, all tumors from the chromosomal instable pathway (CIN) have in common the accumulation of mutations after cell division cycles and loss of chromosomal stability [19]. This pathway is characterized by increased mutation rates, alterations in chromosome number, and rearrangement of chromosomes, which are detected by karyotyping and DNA analysis [19]. Some of the mechanisms by which CIN contributes to CRC tumorigenesis are mutations in key, cell cycle-related genes. These key genes include *BRAF*, *KRAS*, *TP53*, and importantly, the tumor-suppressor *APC* gene which is responsible for familial adenomatous polyposis and 85% of colorectal cancer cases without a hereditary risk factor [20]. In short, the tumor progression of the CIN pathway encompasses the few steps from polyps (adenomas), that can turn normal colorectal epithelium into solid tumors (adenocarcinomas): mutations on the *APC* gene in epithelial cells, mutations in the *KRAS* gene in adenomas, inactivation of the tumor-suppressor gene *TP53* on chromosome 17p, and the deletion of chromosome 18q [20]. Other genes that are closely associated with the functioning of the *APC* gene are also altered in CRC. One example is the *CTNNB1* (β-catenin) as mutations in the *APC* gene result in the overproduction of β-catenin in the cytoplasm and the overactivation of the Wnt signaling pathway [20]. Likewise, mutations in the *KRAS* gene will upregulate the activation of mitogenic pathways such as the mitogen-activated protein kinase (MAPK) pathway, the deleted in the colon cancer (DCC) pathway, and the TGF-β signaling pathway, to name examples [21]. Lastly, mutations in the *TP53* gene are present in 43.28% of CRC cases (mostly missense mutations); they impair the cell's ability to respond to stress, to execute DNA repair, and to arrest cell cycle or implement apoptosis [21].

Following CIN, the second most relevant pathway in CRC is microsatellite instability (MSI). These tumors are characterized by genetic damage of the mismatch repair (MMR) system and they usually present inhibition of the DNA polymerase, that in turn creates short-term insertion-deletion loops (IDL) that further enhance genetic instability [22]. MSI tumors are clinically identified by the abnormal production of the proteins of the DNA MMR system: MLH1, MSH2, MSH6, and PMS2; whose main functions are to repair single base pair mismatches during DNA synthesis and to maintain genomic stability after each cell replication cycle [22]. MSI mostly occurs in the proximal colon and its classification is used as a biomarker for prognosis and standard of care (immunotherapies) based on the 5 microsatellite markers: mononucleotides BAT25 and BAT26, and the dinucleotides D2S123, D5S346, and D17S250 [23]. Moreover, CRC tumors can be classified as MSI-high (MSI-H), those with >30% markers that exhibit genetic instability, MSI-low (MSI-L) for tumors with less than 30% markers with instability or, microsatellite stable (MSS) for tumors with no genetic stability and normal production of the MMR proteins [23]. Contrary to the CIN pathway, MSI tumors are for the most part somatic defects in the MMR genes by either mutational inactivation or by epigenetic silencing of CpG due to aberrant methylation patterns that result in the silencing of promoter genes, although hereditary mutations are frequently reported in *MSH2*, *MSH6* and *PMS2*. In conclusion, MSI pathogenesis arises from the accumulation of mutations and/or epigenetic alteration in several genes rather than the initiation from a single driver [23].

To finalize our genetic factors in the CRC section, we will address the CpG island methylator phenotype (CIMP) serrated pathway. Tumors classified as CIMP have a high number of hyper-methylated genes, with promoters that are silenced and can cause the downregulation of gene expression and protein production [23]. Approximately 35% of CRC cases are CIMP and they have common molecular alterations in other pathways such as the MSI subtype, the hypermethylation of the *MLH1* gene (part of the MMR system) that can lead to DNA repair dysfunction, and the development of hyperplastic polyposis syndrome that also involves the *APC* gene and it is commonly associated with hereditary Lynch syndrome [23]. Overall, CIMP and the serrated CRC tumor subtypes develop faster than other somatic pathways as they combine the rapid accumulation of polyps followed by accumulation of mutations from the defective MMR system and the further silencing of cell cycle genes. Taking all these pathways together, it can be stated that genetic factors in CRC can be hereditary and somatic, with overlapping mutations in key cell cycle genes that contribute to the adenomatous-to-adenocarcinoma transformation in the colon epithelium.

## 5. Tumor biology composition in African American colon cancer patients

To address the specifics of the tumor biology of CRC among African American patients, we would like to start by describing the concept of cancer disparities that are observed in the United States of America (USA). The National Institutes of Health considers the main racial/ethnic minority groups to be African Americans, American Indians and Alaska Natives, Native Hawaiians and Pacific Islanders, and Hispanic/ Latinos [24]. These minority populations are heterogeneous, and we acknowledge that these categories are socially constructed; however, they serve as the official method for tracking cancer incidence, progression, outcomes, and all the other metrics by which cancer care and research is organized in the USA. Hence, this section will start by providing a summary of the cancer disparities that are observed among underrepresented CRC patients and it will continue by focusing on the biological factors that have been studied in African American patients.

There are disparities in the incidence and mortality rates of CRC among all the mentioned minority groups when compared to the average USA population, which is in the majority composed of Caucasian Americans (CA) [24]. For instance, in terms of incidence, it is 45.7 for African American (AA) patients, and 34.1 for Latinos per 100.000 patients. Similar trends are reported for mortality rates, with 19.0 for AA and 11.1 for Latinos, indicating that although progress has been made in the prevention and treatment of CRC in the USA, these patients face challenges that are population-specific [24]. Many factors contribute to this phenomenon, including socioeconomic status and access to healthcare, among others; nevertheless, there are unique biological features that contribute to the tumor biology of CRC in each racial/ ethnic group. In the case of AA and Latinos, it is worth mentioning the disparities in the risk of metastasis due to the reduced prevalence of tumors that are localized and regional (normally cured by surgery or radiation) when compared to CA patients, underscoring the biological differences between these populations [25].

As we discussed previously, CRC results from a combination of the patients' genetic profile (hereditary factors) as well as environmental and somatic changes that will modulate the tumor biology of the colon, such as diet, body mass index, tobacco and alcohol intake, etc. In addition to these contributors, research has highlighted tumor biology that has been associated with each population and their influence in the response to treatment. Studies from clinical trials have shown racial/ethnic disparities in survival rates of stage III CRC cancer even when patients received the same standard of care, greater toxicity in response to 5-fluorouracil (5-FU) therapy regimens, and unique pharmacogenetic variants in AA patients [26]. Also, the frequency of the MSI/MMR-deficient tumor subset, which is associated with better prognosis and serves as a biomarker for immunotherapies, appears to be reduced in AA when compared to CA patients (14 in CA vs. 7% in AA) [27].

Regarding the inherited, germline-associated syndromes in CRC, such as the familial adenomatous polyposis, AA seem to have a prevalence comparable to the other populations in the USA [28]. Despite that, previous investigations have identified unique somatic mutations in the 5-Hydroxytryptamine Receptor 1F (*HTR1F*), Folliculin (*FLCN*), and EPH Receptor A6 (*EPHA6*) protein-genes that in AA colon cancer patients seem to play a role in worse prognosis and greater chemoresistance [29]. These novel somatic mutations in AA suggest that *EPHA6* and *FLCN* could serve as driver genes for CRC in these patients and they pinpoint the need for genetic studies that target specific populations. One of these studies demonstrated that the secretion of the interleukin-6 cytokine and oxidative species in the colon epithelium could impair the functioning of the MSH3 protein (MMR system) and promote the development of MSI-L, elevated microsatellite alterations at selected tetranucleotide repeats (EMAST). EMAST is a subtype that is more prevalent in AA patients when compared to CA [30]. These findings, in correlation with the reported lower rates of MSI tumors among AA patients, may encourage researchers to further investigate the influence of inflammation and the intersection of the immune system with the tumor biology of CRC in AA patients.

## 6. Role of miRNAs in colon cancer and African Americans

microRNAs (miRNAs), powerful regulatory RNAs 18–24 nt in length, play a major role in oncology pathways. As miRNAs can potentially serve as biomarkers for CRC at several levels, understanding miRNA dysregulation is important in defining its effects on biomolecular pathways and developing possible therapeutic targets. This section will discuss miRNA expression patterns characteristic of CRC tumors, miRNAs identified as potential non-invasive biomarkers, their role in chemo-response, their contribution to racial health disparities, and their use as therapeutic targets.

miRNA dysregulation is a common component of many cancers, including in CRC. Among the hundreds of discovered miRNAs, the most relevant miRNAs unique to CRC tumors compared to healthy tissue, as well as their primary oncological function, is summarized in **Table 1** (upregulated miRNAs) and **Table 2** (downregulated miRNAs). Notably, available data from The Cancer Genome Atlas was used to provide

| **[CRC tumor miRNA profile] summary of upregulated miRNAs (tumor vs. normal tissue)** | | |
|---|---|---|
| **Biological function** | **Upregulated miRNAs** | **Reference** |
| Promotes cell cycle/proliferation | 17-3p, 20a, 21, 26a, 31, 106a, 135a/b, 141, 200c, 301a, 598, 1273g-3p | [31–42] |
| Promotes migration/invasion/metastasis | 20a, 21, 26a, 29a, 31, 135a/b, 155, 200c, 224, 301a, 494, 1273g-30 | [32–35, 37, 39, 40, 42–46] |
| Inhibits apoptosis | 17-3p, 31, 92a, 106a, 135a/b, 200c | [31, 35–37, 39, 47] |
| Involved in drug sensitivity/resistance | 96, 155, 192/215 | [44, 48, 49] |
| Hypoxia/ROS regulation | 210 | [50] |
| Malignant transformation | 182/503 | [51] |
| CRC stem cell tumorigenicity | 221 | [52] |
| Ambiguous | 18a (both an oncomiR and Tumor Suppressor), 217 (inhibits proliferation/promotes apoptosis) | [53, 54] |

**Table 1.**
*Upregulated miRNAs in CRC tumor tissue vs. normal tissue.*

| [CRC tumor miRNA profile] summary of downregulated miRNAs (tumor vs. normal tissue) | | |
|---|---|---|
| **Biological function** | **Downregulated miRNAs** | **Reference** |
| Inhibits cell cycle/proliferation | 7, 18a-3p, 27b, 30a, 101, 125a/b, 126, 143/145, 144, 149, 155, 186-5p, 194, 205-5p, 216a-3p, 320a, 330, 374b, 375, 383, 455, 486, 511, 744, 1271, let-7 | [53, 55–79] |
| Inhibits migration/invasion/metastasis | 7, 19b-1, 26b, 27b, 101, 125a/b, 126, 155, 186-5p, 205-5p, 320a, 328, 330, 374b, 455, 511, 744, 1271 | [55, 56, 58–60, 64, 65, 67, 69–71, 74, 76–78, 80–82] |
| Promotes apoptosis | 7, 18a-3p, 30a, 101, 129, 143/145, 149, 155, 455, 511, 744, 1271, let-7 | [53, 55, 57, 59, 63, 64, 74, 76–79, 83] |
| Involved in drug sensitivity/resistance | 26b, 328, 1271 | [78, 81, 82, 84] |
| Involved in angiogenesis | 375 | [72] |

**Table 2.**
*Downregulated miRNAs in CRC tumor tissue vs. normal tissue.*

a comparative analysis of miRNA expressed in colon tumors (N = 253) versus uninvolved normal tissues (N = 8). Of these, 39 upregulated and 54 downregulated miRNAs were implicated in colon cancer and, 9 of them were critical with a downstream impact on 461 genes associated with patient survival [85]. Among others, pathways affected by these miRNAs include Wnt signaling, p53, cell adhesion, cAMP signaling, stem cell pluripotency, MAPK, and HIF-1 [85]. In the pathway network of these miRNAs, five 'hub' genes (genes that have high connectivity to oncological pathways) were identified as mediating the function of these miRNAs: *PPARGC1A*, *COLIA1*, *SYT1*, *PGR*, and *KCNB1* [85]. Additionally, a study of patients with stage III CRC showed that 11 miRNAs (miR-135b, miR-141, miR-18a, miR-20a, miR-21, miR-224, miR-29a, miR-31, miR-34a, miR-92a, and miR-96) were overexpressed in tumors relative to their matching normal samples [86]. In addition to deletion and mutation, hypermethylation of certain miRNA promoters contributes to the increased dysregulation of miRNAs in CRC [87].

The feasibility of using miRNAs as biomarkers stems from their general stability in blood, owing to their structure which resists RNAse-mediated degradation [88]. miRNAs that may serve as CRC non-invasive biomarkers are summarized in **Table 3** [101]. Thus, exosomes-circulating miRNAs may have value as early, non-invasive CRC biomarkers [89]. For example, serum from CRC patients contained 69 miRNAs that were significantly upregulated compared to normal subjects. Additionally, cell culture media from five CRC cell lines demonstrated 52 upregulated miRNAs compared to culture media from normal colon epithelial cells. For both these *in vivo* and *in vitro* data, 16 miRNA (let-7a, miR-1224-5p, miR-1229, miR-1246, miR-1268, miR-1290, miR-1308, miR-150, miR-181b, miR-181d, miR-1915, miR-21, miR-223, miR-23a, miR-483-5p, and miR-638) were commonly upregulated compared to a non-cancerous baseline. Following surgical resection of tumors in 29 CRC patients of all stages (I-IV), serum levels of eight biomarker miRNAs (let-7a, miR-1229, -1246, -1224-5p, -150, -21, -223, -23a) were significantly decreased compared to pre-resection levels [89]. Importantly, significantly higher serum levels of miR-18a and miR-29a were observed in CRC patients when compared to levels in healthy individuals as control [86]. Therefore, circulating miRNAs have emerged as potential non-invasive predictive biomarkers for CRC.

| [Diagnostic biomarkers] summary of upregulated in serum/plasma (CRC vs. healthy patient) | | |
|---|---|---|
| **Location of biomarker** | **miRNA** | **Reference** |
| Upregulated in serum (CRC patient vs. healthy control) **AND** culture media (CRC cell lines, n = 5, vs. colon epithelial cells) | 21, 23a, 150, 181b/d, 223, 483-5p, 638, 1224-5p, 1229, 1246, 1268, 1290, 1308, 1915, let-7A | [89] |
| Upregulated in serum (CRC patients vs. healthy control) | 18a, 24, 24-2, 29a, 122, 135a-5p, 139-3p, 139-5p, 203, 320a, 423-5p, 6826 | [86, 90–97] |
| Markers in plasma exosomes | 17-5p, 21, 92a-3p, 6803-5p | [98–100] |

**Table 3.**
*Summary of miRNAs that have potential as non-invasive biomarkers.*

Since miRNAs modulate drug targets directly or indirectly, response to standard of care (SOC) chemotherapeutic agents may be influenced by the dysregulation of select miRNAs. Indeed, expression of miRNAs is altered upon treatment of CRC cell lines with 5-fluorouracil (5-FU). Patients who did not respond well to fluoropyrimidine chemotherapy had higher plasma levels of miR-106, miR-484, and miR-130b [102]. Furthermore, higher miR-27B, miR-148A, and miR-326 levels were associated with decreased progression-free survival, whereas miR-326 was related to decreased overall survival [102]. Additionally, 5-FU reduces miR-200b, which lowers the levels of the protein tyrosine phosphatase, PTPN12, which, in turn, downregulates oncogenes, including *c-ABL* and *RAS*, resulting in decreased cell proliferation [103]. Further assessment of these miRNAs in drug uptake and metabolism will help characterize their significance in chemotherapeutic response and pave the way for more personalized treatment plans.

Lacking in our understanding, due to the absence of or limited use of AA samples in bench and clinical studies, is the role of miRNA in CRC racial health disparity in terms of cancer initiation and chemo-response. Microarray and qPCR analysis implicated miR-182, -152, -204, -222, and -202 when comparing AA and CA tumor samples [104]. Among these, miR-182 was the most significant—upregulated in AA vs. CA—in race tumor interactions. *FOXO1* and *FOXO3A*, miR-182 targets, were shown to be downregulated in AA colon tumors compared to the colon tumors of CAs [104]. Findings by Bovell and colleagues "suggest that the prognostic value of miRNAs in colorectal cancers varies with patient race/ethnicity and stage of disease [105]. 5 miRNAs (*miR-20a, -21, -106a, -181b*, and *-203*) in paired normal and tumor CRC had higher expression in CRC than in adjacent non-involved tissues. High expression of *miR-203* was associated with poor survival of CA patients with stage IV CRC and with poor survival of AA patients with stages I and II colorectal cancers. High expression of *miR-21* and *miR-181b* correlated with poor survival of CA (stage IV) and AA (stage III) patients, respectively. These analyses suggest that a deeper biomolecular understanding of miRNA dysregulation between racial and ethnic groups may provide a richer context in addressing the clinically observed health disparity through more personalized treatments.

The ability of miRNA to regulate expression of many downstream genes and the proficiency of biotechnology to synthesize oligonucleotides, promote and enable the use of these molecules as potential therapeutic agents. miR-34a, a central miRNA in the p53 stress pathway, is often lost in CRC and has been the hallmark of miRNA mimic therapy. In a phase 1 clinical trial, liposomal miR-34a mimics were shown to provide benefits against advanced solid tumors; however, these mimics were accompanied by many off-target-side-effects and adverse immune reactions [106]. Other strategies may include inhibiting oncogenic miRNA or more specific targeting of tumor miRNA replacement therapy. The challenge of miRNAs as therapeutic agents is a limited understanding of their targeted pathways in various tissues.

miRNAs have become vital as prognostic and therapeutic biomarkers/targets for cancer. miRNA dysregulation in CRC, along with their role in racial health disparities, is continually being explored. Identification and profiling of miRNAs in diverse patient population will result in the generation of personalized therapeutic targets which will allow for optimal patient care.

## 7. Methylation patterns and alterations in colon cancer in African Americans

A major component of gene expression is DNA methylation. It is well documented that aberrant methylation patterns in CRC contribute to tumorigenesis and progression. This section will discuss dysregulated methylation patterns of CRC; specifically, pathways affected by aberrant methylation (hypo−/hypermethylation) and differential methylation of CRC tumors of African American patients.

In general, 10–40% of CRC tumor cells have a hypomethylated genome [107]. Most of this hypomethylation occurs in repetitive elements and influences the initiation of tumorigenesis [108]. *In vivo*, mice possessing a knockout of DNA methyltransferase displayed increased genomic instability and tumor initiation [109]. While methylation can alter gene promoters, methylation of non-coding regions like long interspersed nuclear elements (LINE) can also affect adjacent gene expression. In addition, hypomethylation of regulatory elements may lead to unregulated oncogene expression [110]. It is important to mention, however, that methylation of a gene/gene promoter is not necessarily a guarantee of decreased gene mRNA expression in the cell and that more complex downstream mechanisms may be at play [111].

In addition to hypomethylation, specific promoters involved in CRC can be hypermethylated. While hypermethylation of tumor suppressors is a normal part of aging, some methylation patterns describe preferential hypermethylation of tumor suppressors [112]. For example, the CpG-island methylator phenotype (CIMP) produces a subtype of CRC in which CpG islands of tumor suppressor genes become hypermethylated through an epigenetic instability pathway [112]. As proposed by Ehrlich and colleagues, the consequences of aberrant methylation include genomic instability, epigenetic inactivation of tumor suppressors, altered chromatin heterostructure interactions, and activation of oncogenic elements [107].

Specific CRC-relevant pathways are affected by methylation. A few are described here. In Wnt signaling, the *APC* promoter is methylated in about 18% of CRCs. However, this does not necessarily correlate with a subsequent decrease in *APC* expression of downstream target expression [113]. Furthermore, methylation of Wnt inhibitors (*SFRP1*, 2, 4, and 5) is an early event common in CRC [114]. Wnt signaling malfunction is an early hallmark and driver of CRC progression [115]. Therefore, if intrinsic inhibitors of Wnt signaling, like *SFRP*s, are downregulated via methylation, the cell may have a propensity for aberrant upregulated Wnt signaling [114]. In the p53 pathway, while direct methylation of *TP53* is rarely observed, components like p14-ARF (which sequesters MDM2 ubiquitin ligase) have been methylated and downregulated in 20% of microsatellite instable (MSI) CRCs [116]. For the RAS pathway, up to 80% of CRCs have *RASSF1/2* promoters methylated. These proapoptotic gene products are modulators of the RAS pathway. Thus, RASSF1/2 downregulation may promote tumorigenesis [117]. Finally, TSP-1—an extracellular matrix glycoprotein—cleaves TFGβ into the active form. This gene is often methylated at its gene promoter in about 20% of CRCs [118].

Aberrant methylation in CRC has been shown to stem from several sources. First, deregulation of relevant methylation enzymes like DNA methyltransferases (DMNT) can kickstart aberrant methylation. While DMNT is rarely mutated in CRC (unlike other cancers), the protein is overexpressed but not related to a specific aberrant methylation phenotype [119, 120]. *TET1* methylation has been suggested in the progression of CIMP CRC [121]. Mechanisms that normally protect the genome from aberrant methylation (e.g., DNA-binding proteins, RNA polymerase, or histone binding) may be modified which allows nearby hypermethylated regions to affect previously nonmethylated areas [122]. Specifically, Turker proposes a hypothesis wherein long-term methylation of nearby promoters initiates due to a constantly shifting methylation 'boundary' of adjacent hypermethylated regions [122]. As an unmethylated gene promoter couples and decouples with transcription factors, the 'boundary' of the adjacent hypermethylated region is in flux and can thus spontaneously/iteratively spread towards nearby CpG islands in the gene promoter, ultimately favoring a gene-repressing DNA superstructure [122]. Additionally, in response to oxidative stress, the DNA damage repair system recruits DMNTs which are implicated in methylation of nearby promoters [123]. Currently, DNMT inhibitors are being examined for use as an adjunct therapy to canonical chemotherapeutics for CRC; specifically, for prevention of the hypermethylation of tumor suppressor genes. These therapies have had some success *in vitro* and in mouse models, but not in clinical trials [112].

Importantly, it was reported in one study that the CRC tumors of AA patients had 14.6-fold more hypermethylated regions and 25-fold more hypomethylated regions than CA in respect to tumor versus normal tissue [124]. In AA tumors, *CHL1*, four inflammatory genes (*NELL1*, *GDF1*, *ARHGEF4*, and *ITGA4*), and 7 miRNAs were methylated. Of these miRNAs, miR-9-3P and miR-124-3P are implicated in CRC while the targets of miR-124 (which was hypermethylated) were upregulated in AA vs. CA [124]. In a separate study, four methylation target genes were observed in AA samples: *BPM3*, *EID3*, *GAS7*, and *GPR75* [110]. Differential methylation patterns in different groups of patients may inform more efficacious personalized treatment protocols, particularly in the field of DMNT inhibitors.

In addition to mutations, dysregulation of gene expression through epigenetic alterations (i.e., methylation) impacts the initiation and progression of CRC. However, while global hypomethylation and local hypermethylation are prevalent, more specific patterns may need to be analyzed for therapeutic consideration. Beyond the comparisons of CRC tumors to normal tissue concerning methylation patterns, although limited in the scope of race and ethnicity, there are also marked differences between tumors originating from different racial groups. However, these findings may be instrumental in predicting tumor aggressiveness and responsiveness to standard of care and novel treatment modalities.

## 8. Immunological profiles of colon cancer in African Americans

As we discussed in the introduction, evasion of immune surveillance is one of the hallmarks of cancer cells. Hence, we will discuss how genetic factors, tumor biology, and genetic regulators (miRNAs and DNA methylation patterns) intersect with the immune system in CRC in AA patients.

Colon tumors closely interact with immune cells that reside at the tumor site (microenvironment) and immune cells that are part of the systemic immune surveillance system. They both play a role in tumor progression, prognosis, and response to treatment in CRC [125]. Lymphocytes, cytotoxic $CD8^+$ T cells to be precise, are one of the most efficient immune cells to perform surveillance, to limit the tumor progression in CRC and their filtration into tumors and are associated with better prognosis and outcome. The presence of high levels of $CD8^+$ T cells in the center and invasive margins of colon tumors have been associated with improved patients' survival when compared with patients with low infiltration of the same cell type [126]. These lymphocytes can ignite apoptosis in target cells by the secretion of (among others) the serine protease Granzyme B. Granzyme $B^+$ T and Natural Killer (NK) cells are activated in response to the presence of neoantigens in the surface of cancer cells (recognized by antigen-presenting cells); especially from hypermutated tumors that are mostly classified as MSI due to their genetic instability and MMR deficiency [126]. In the context of AA CRC patients, however, it has been demonstrated that in MSI-H tumors, these patients presented lower infiltration of $CD8^+$ T cells when compared to tumors from the same subtype from CA patients [127]. Furthermore, a study that investigated 250 CRC cases and compared MSS tumors from AA and CA patients found that tumors from AA patients had lower numbers of GRANZYME $B^+$ lymphocytes, suggesting that CRC in AA is characterized by impaired immune surveillance and lower cytotoxicity regardless of tumor type [128].

These disparities in the immunological profile of CRC tumors in AA patients also influence the access of these patients to the immunotherapies available for CRC cancer. For example, when cytotoxic T cells are activated by tumoral antigens they will induce memory T cells that are characterized by the expression of the programmed cell death protein 1 (PD-1) receptor on their surface that serves as a negative feedback loop for the inactivation of these lymphocytes and prevents auto-immunity [129]. This receptor will interact with the PD-1 ligand that can be on the surface of cancer cells as a tumoral strategy of immune-surveillance evasion, and it is the target of the PD-1/PD-L1 antibody therapy that blocks this interaction and releases $CD8^+$ T cells from the negative effects of the PD-L1 ligand. Not surprisingly, and as we mentioned in the tumor biology section, the MSI tumor classification is a biomarker for access to this immunotherapy based on the direct correlation of hypermutation in cancer cells, expression of neo-antigens and PD-L1 ligands, response to the anti-PD-1 antibody therapy, and positive outcome [130].

Remarkably, a research study that compared the gene expression of CRC tumors from AA and CA patients confirmed that tumors from AA had a lower expression of the *GZMB* gene (which codes for the GRANZYME B protein) and lower expression of *PDL1* (gene encoding for the PDL1 ligand), results that correlate the previously described findings at the protein and cellular levels [131]. Furthermore, this investigation demonstrated that colon tumors from AA presented significantly higher numbers of exhausted (or functionally impaired) $CD8^+$ T cells and instead, had higher numbers of pro-inflammatory myeloid cells that are associated with chronic inflammation and worse prognosis. Lastly, the authors measured the levels of T cell-related cytokines in the plasma from both cohorts and demonstrated that CA had significantly higher levels of interleukin 12 and other $CD8^+$ T cells-activating cytokines when compared to AA, proposing that the immunological profile of these patients present disparities at the tumor site and at the systemic level [131]. The combined conclusions from these studies in AA patients demonstrated that there is a lower incidence of MSI-H tumors in this population, lower infiltrations of cytotoxic T cells in MSI-H and MSS tumor subtypes, and a reduced gene expression and activation of GRANZYME B$^+$ cells. Taken together, these research studies indicate that AA patients present an impaired immunosurveillance mechanism and may have lower access to the *PD-1/PD-L1* immunotherapy for CRC cancer.

## 9. Future directions of gene expression and colon cancer in African Americans

The push for research in CRC racial health disparity and inclusivity in clinical trials represents a major step forward in personalized medicine and optimizing health outcomes for a diverse patient population. To continue the acceleration of current progress, new future directions in the analysis and development of novel *in vitro* and *in vivo* models are necessary. In terms of analytical frameworks, multi-omics assessment of CRC will help unlock a new level of comprehension and possible treatment pathways. These types of analyses, often including genomic, epigenetic, and expression analyses, can reveal a richer set of conclusions and holistic understanding of the disease and underlying racial health disparity [132]. This perspective can be elaborated when considering synergy with other physiological analyses. Intersectionality between gene expression and metabolomics studies can supply a detailed description of compound effects and mechanisms. For example, studies with the compound shikonin were analyzed with CRC through integrated transcriptome and metabolomic perspectives which can allow for more robust predictive conclusions [133]. Mouse model studies demonstrate the ability of microbiome dysbiosis to epigenetically influence gene expression profiles of the colonic epithelium towards more inflammatory or CRC risk patterns [134].

Not only will these multidimensional analyses aid the progress of CRC research, but so will the development of tools that seek to make racial health disparity research more accessible. Social determinants of health are important components of health disparity; however, collective studies have demonstrated that research observations support differences in the distribution and pattern of driver mutations in diseases such as colon cancer that present more in AA patients as compared to CA patients. The etiologic basis for differences between race groups, such as higher rates of *KRAS* mutant tumors in AA colon cancer patients, is not known, and relevance to tumor behavior including effect on chemotherapy responsiveness remains unclear. Importantly, the lack of cell lines, organoids, and/or patient-derived xenograft models representative of disease heterogeneity that reflect differences in disease patterns by race severely limits our understanding of and ability to study the differences in disease behavior between patient populations based on race. This includes assessments of differences in treatment response where much of what is known is based on a few models from CA patients. Overall, the limited availability of racially diverse tissue for research purposes and the lack of therapeutic models hampers the ability to evaluate cancer initiation, progression, and therapy in an inclusive population. This one factor contributes most significantly to gaps in the knowledge of cancer in racially and ethnically distinct populations. This should be seen as a scientific area of high priority needed to reduce the unequal burden of cancer health disparities. Thus, what is extremely important is a more concerted effort to generate diverse *in vitro*, *ex vivo*, and *in vivo* models. To date, the American Type Culture Collection (ATCC, founded

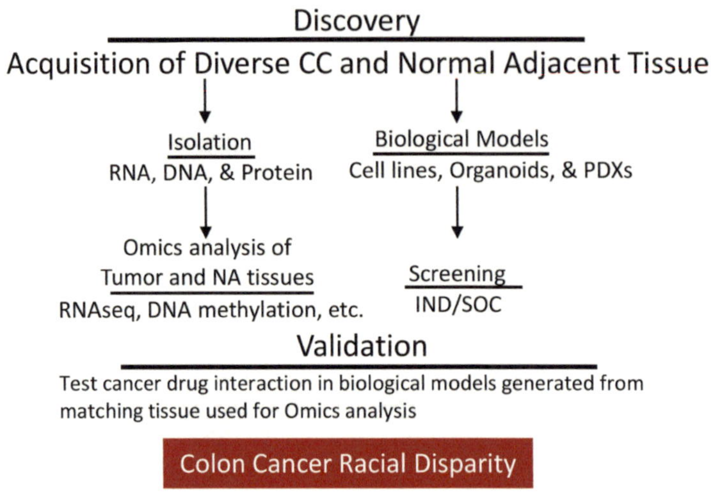


**Figure 2.**
*Assessment methods to address differences in response to treatment.*

in 1925) has a limited number of cell lines designated as AA or Hispanic American (HA), some organ tumors fair better than other organ tumors. For example, in the ATCC repository, there are no colon cancer cell lines designated as AA. Equally important is the correlation of tumor biology discovery with metadata linking social determinates of health. To meet this need, the establishment of three novel AA CRC cell lines now allows for *in vitro* and *in vivo* assessment of compounds and tumor biology which may influence differential chemo-response by race [135]. 3D cell culture in Matrigel©-related protocols may further elucidate these tools for more realistic *in vitro* experimentation [135]. Furthermore, our research group and others are developing colon tumor organoids to better recapitulate chemotherapeutic responses representative of the natural tumor environment. Finally, patient-derived xenografts from tumor tissue resected from African American CRC patients are a valuable source of cells for downstream applications like cell culture, primary cell lines, organoids, chemotherapeutic assessment, etc.

Multi-omics assessment combined with accessible tools, the exploration of CRC gene expression, and overall biology will make possible an understanding that can be therapeutically targeted for the maximum benefit of the patient. Discovery and validation methods are provided in **Figure 2**. Given the multidimensionality of CRC, it is scarcely sustainable to maintain blanket standards of care that have abhorrent side effects with a sizeable chance of failure. Ideally, sampling a patient's tumor and having a pipeline of treatments optimized for their genetic, epigenetic, and metabolomic profile will be the future of cancer treatment. In a diverse patient population with racial health disparities among other socioeconomic obstacles, research aimed at unpacking these complexities is vital for getting closer to the goal of a personalized approach for healing patients.

## Author details

Jennie Williams[1*], Jenny Paredes[2] and Shrey Thaker[1]

1 Stony Brook University, Stony Brook, USA

2 Memorial Sloan Kettering Cancer Center, New York, USA

*Address all correspondence to: jennie.williams@stonybrookmedicine.edu

## References


[1] What is Cancer: National Cancer Institute. 2021. Available from: https://www.cancer.gov/about-cancer/understanding/what-is-cancer

[2] Weinberg RA. Coming full circle-from endless complexity to simplicity and back again. Cell. 2014;**157**(1):267-271

[3] Horne SD, Pollick SA, Heng HH. Evolutionary mechanism unifies the hallmarks of cancer. International Journal of Cancer. 2015;**136**(9):2012-2021

[4] Witsch E, Sela M, Yarden Y. Roles for growth factors in cancer progression. Physiology (Bethesda, Md.). 2010; **25**(2):85-101

[5] Pylayeva-Gupta Y, Grabocka E, Bar-Sagi D. RAS oncogenes: Weaving a tumorigenic web. Nature Reviews. Cancer. 2011;**11**(11):761-774

[6] Stracquadanio G, Wang X, Wallace MD, Grawenda AM, Zhang P, Hewitt J, et al. The importance of p53 pathway genetics in inherited and somatic cancer genomes. Nature Reviews. Cancer. 2016;**16**(4):251-265

[7] Cheung KJ, Ewald AJ. A collective route to metastasis: Seeding by tumor cell clusters. Science. 2016;**352**(6282):167-169

[8] Ye X, Weinberg RA. Epithelial-mesenchymal plasticity: A central regulator of cancer progression. Trends in Cell Biology. 2015;**25**(11):675-686

[9] DeBerardinis RJ, Chandel NS. Fundamentals of cancer metabolism. Science Advances. 2016;**2**(5):e1600200

[10] Pavlova NN, Thompson CB. The emerging hallmarks of cancer metabolism. Cell Metabolism. 2016;**23**(1):27-47

[11] DeBerardinis RJ, Cheng T. Q's next: The diverse functions of glutamine in metabolism, cell biology and cancer. Oncogene. 2010;**29**(3):313-324

[12] Junttila MR, de Sauvage FJ. Influence of tumour micro-environment heterogeneity on therapeutic response. Nature. 2013;**501**(7467):346-354

[13] Dunn GP, Bruce AT, Ikeda H, Old LJ, Schreiber RD. Cancer immunoediting: From immunosurveillance to tumor escape. Nature Immunology. 2002;**3**(11):991-998

[14] Malmberg KJ. Effective immunotherapy against cancer: A question of overcoming immune suppression and immune escape? Cancer Immunology, Immunotherapy. 2004;**53**(10):879-892

[15] Arnold M, Sierra MS, Laversanne M, Soerjomataram I, Jemal A, Bray F. Global patterns and trends in colorectal cancer incidence and mortality. Gut. 2017;**66**(4):683-691

[16] Jasperson KW, Tuohy TM, Neklason DW, Burt RW. Hereditary and familial colon cancer. Gastroenterology. 2010;**138**(6):2044-2058

[17] Fleming M, Ravula S, Tatishchev SF, Wang HL. Colorectal carcinoma: Pathologic aspects. Journal of Gastrointestinal Oncology. 2012; **3**(3):153-173

[18] Keum N, Giovannucci E. Global burden of colorectal cancer: Emerging trends, risk factors and prevention strategies. Nature Reviews.

Gastroenterology & Hepatology. 2019;**16**(12):713-732

[19] Lao VV, Grady WM. Epigenetics and colorectal cancer. Nature Reviews. Gastroenterology & Hepatology. 2011;**8**(12):686-700

[20] Pino MS, Chung DC. The chromosomal instability pathway in colon cancer. Gastroenterology. 2010;**138**(6):2059-2072

[21] Nguyen HT, Duong HQ. The molecular characteristics of colorectal cancer: Implications for diagnosis and therapy. Oncology Letters. 2018;**16**(1):9-18

[22] Boland CR, Goel A. Microsatellite instability in colorectal cancer. Gastroenterology. 2010;**138**(6):2073-87.e3

[23] Koveitypour Z, Panahi F, Vakilian M, Peymani M, Seyed Forootan F, Nasr Esfahani MH, et al. Signaling pathways involved in colorectal cancer progression. Cell & Bioscience. 2019;**9**:97

[24] Cancer Disparities cancer.gov: National Cancer Institute. 2020. Available from: https://www.cancer.gov/about-cancer/understanding/disparities

[25] Lansdorp-Vogelaar I, Kuntz KM, Knudsen AB, van Ballegooijen M, Zauber AG, Jemal A. Contribution of screening and survival differences to racial disparities in colorectal cancer rates. Cancer Epidemiology, Biomarkers & Prevention. 2012;**21**(5):728-736

[26] Dimou A, Syrigos KN, Saif MW. Disparities in colorectal cancer in African-Americans vs whites: Before and after diagnosis. World Journal of Gastroenterology. 2009;**15**(30):3734-3743

[27] Ashktorab H, Ahuja S, Kannan L, Llor X, Ellis NA, Xicola RM, et al. A meta-analysis of MSI frequency and race in colorectal cancer. Oncotarget. 2016;**7**(23):34546-34557

[28] Carethers JM. Racial and ethnic factors in the genetic pathogenesis of colorectal cancer. Journal of the Association for Academic Minority Physicians. 1999;**10**(3):59-67

[29] Guda K, Veigl ML, Varadan V, Nosrati A, Ravi L, Lutterbaugh J, et al. Novel recurrently mutated genes in African American colon cancers. Proceedings of the National Academy of Sciences of the United States of America. 2015;**112**(4):1149-1154

[30] Koi M, Tseng-Rogenski SS, Carethers JM. Inflammation-associated microsatellite alterations: Mechanisms and significance in the prognosis of patients with colorectal cancer. World Journal of Gastrointestinal Oncology. 2018;**10**(1):1-14

[31] Lu D, Tang L, Zhuang Y, Zhao P. miR-17-3P regulates the proliferation and survival of colon cancer cells by targeting Par4. Molecular Medicine Reports. 2018;**17**(1):618-623

[32] Zhu GF, Xu YW, Li J, Niu HL, Ma WX, Xu J, et al. Mir20a/106a-WTX axis regulates RhoGDIa/CDC42 signaling and colon cancer progression. Nature Communications. 2019;**10**(1):112

[33] Zhao J, Zhang Y, Zhao G. Emerging role of microRNA-21 in colorectal cancer. Cancer Biomarkers. 2015;**15**(3):219-226

[34] Coronel-Hernández J, López-Urrutia E, Contreras-Romero C, Delgado-Waldo I, Figueroa-González G, Campos-Parra AD, et al. Cell migration and proliferation are regulated by miR-26a in colorectal cancer via the PTEN-AKT axis. Cancer Cell International. 2019;**19**:80

[35] Yang X, Xu X, Zhu J, Zhang S, Wu Y, Wu Y, et al. miR-31 affects colorectal cancer cells by inhibiting autophagy in cancer-associated fibroblasts. Oncotarget. 2016;**7**(48):79617-79628

[36] He Y, Wang G, Zhang L, Zhai C, Zhang J, Zhao X, et al. Biological effects and clinical characteristics of microRNA-106a in human colorectal cancer. Oncology Letters. 2017;**14**(1):830-836

[37] Valeri N, Braconi C, Gasparini P, Murgia C, Lampis A, Paulus-Hock V, et al. MicroRNA-135b promotes cancer progression by acting as a downstream effector of oncogenic pathways in colon cancer. Cancer Cell. 2014;**25**(4):469-483

[38] Ding L, Yu LL, Han N, Zhang BT. miR-141 promotes colon cancer cell proliferation by inhibiting MAP2K4. Oncology Letters. 2017;**13**(3):1665-1671

[39] Chen J, Wang W, Zhang Y, Hu T, Chen Y. The roles of miR-200c in colon cancer and associated molecular mechanisms. Tumour Biology. 2014;**35**(7):6475-6483

[40] Fang Y, Sun B, Xiang J, Chen Z. MiR-301a promotes colorectal cancer cell growth and invasion by directly targeting SOCS6. Cellular Physiology and Biochemistry. 2015;**35**(1):227-236

[41] Li KP, Fang YP, Liao JQ, Duan JD, Feng LG, Luo XZ, et al. Upregulation of miR-598 promotes cell proliferation and cell cycle progression in human colorectal carcinoma by suppressing INPP5E expression. Molecular Medicine Reports. 2018;**17**(2):2991-2997

[42] Li M, Qian X, Zhu M, Li A, Fang M, Zhu Y, et al. miR-1273g-3p promotes proliferation, migration and invasion of LoVo cells via cannabinoid receptor 1 through activation of ERBB4/PIK3R3/mTOR/S6K2 signaling pathway. Molecular Medicine Reports. 2018;**17**(3):4619-4626

[43] Tang W, Zhu Y, Gao J, Fu J, Liu C, Liu Y, et al. MicroRNA-29a promotes colorectal cancer metastasis by regulating matrix metalloproteinase 2 and E-cadherin via KLF4. British Journal of Cancer. 2014;**110**(2):450-458

[44] Gao Y, Liu Z, Ding Z, Hou S, Li J, Jiang K. MicroRNA-155 increases colon cancer chemoresistance to cisplatin by targeting forkhead box O3. Oncology Letters. 2018;**15**(4):4781-4788

[45] Fassan M, Cui R, Gasparini P, Mescoli C, Guzzardo V, Vicentini C, et al. miR-224 is significantly upregulated and targets Caspase-3 and Caspase-7 during colorectal carcinogenesis. Translational Oncology. 2019;**12**(2):282-291

[46] Sun HB, Chen X, Ji H, Wu T, Lu HW, Zhang Y, et al. miR-494 is an independent prognostic factor and promotes cell migration and invasion in colorectal cancer by directly targeting PTEN. International Journal of Oncology. 2014;**45**(6):2486-2494

[47] Tsuchida A, Ohno S, Wu W, Borjigin N, Fujita K, Aoki T, et al. miR-92 is a key oncogenic component of the miR-17-92 cluster in colon cancer. Cancer Science. 2011;**102**(12):2264-2271

[48] Ge T, Xiang P, Mao H, Tang S, Zhou J, Zhang Y. Inhibition of miR-96 enhances the sensitivity of colorectal cancer cells to oxaliplatin by targeting TPM1. Experimental and Therapeutic Medicine. 2020;**20**(3):2134-2140

[49] Boni V, Bitarte N, Cristobal I, Zarate R, Rodriguez J, Maiello E, et al. miR-192/miR-215 influence 5-fluorouracil resistance through cell cycle-mediated mechanisms complementary to its post-transcriptional thymidilate

synthase regulation. Molecular Cancer Therapeutics. 2010;**9**(8):2265-2275

[50] Ullmann P, Qureshi-Baig K, Rodriguez F, Ginolhac A, Nonnenmacher Y, Ternes D, et al. Hypoxia-responsive miR-210 promotes self-renewal capacity of colon tumor-initiating cells by repressing ISCU and by inducing lactate production. Oncotarget. 2016;**7**(40):65454-65470

[51] Li L, Sarver AL, Khatri R, Hajeri PB, Kamenev I, French AJ, et al. Sequential expression of miR-182 and miR-503 cooperatively targets FBXW7, contributing to the malignant transformation of colon adenoma to adenocarcinoma. The Journal of Pathology. 2014;**234**(4):488-501

[52] Mukohyama J, Isobe T, Hu Q, Hayashi T, Watanabe T, Maeda M, et al. miR-221 targets QKI to enhance the tumorigenic capacity of human colorectal cancer stem cells. Cancer Research. 2019;**79**(20):5151-5158

[53] Kolenda T, Guglas K, Kopczyńska M, Sobocińska J, Teresiak A, Bliźniak R, et al. Good or not good: Role of miR-18a in cancer biology. Reports of Practical Oncology and Radiotherapy. 2020;**25**(5):808-819

[54] Wang B, Shen ZL, Jiang KW, Zhao G, Wang CY, Yan YC, et al. MicroRNA-217 functions as a prognosis predictor and inhibits colorectal cancer cell proliferation and invasion via an AEG-1 dependent mechanism. BMC Cancer. 2015;**15**:437

[55] Xu K, Chen Z, Qin C, Song X. miR-7 inhibits colorectal cancer cell proliferation and induces apoptosis by targeting XRCC2. Oncotargets and Therapy. 2014;7:325-332

[56] Luo Y, Yu SY, Chen JJ, Qin J, Qiu YE, Zhong M, et al. MiR-27b directly targets Rab3D to inhibit the malignant phenotype in colorectal cancer. Oncotarget. 2018;**9**(3):3830-3841

[57] Xie M, Qin H, Luo Q, Huang Q, He X, Yang Z, et al. MicroRNA-30a regulates cell proliferation and tumor growth of colorectal cancer by targeting CD73. BMC Cancer. 2017;**17**(1):305

[58] Yang M, Tang X, Wang Z, Wu X, Tang D, Wang D. miR-125 inhibits colorectal cancer proliferation and invasion by targeting TAZ. Bioscience Reports. 2019;**39**(12)

[59] Yang Q, Yu W, Han X. Overexpression of microRNA-101 causes anti-tumor effects by targeting CREB1 in colon cancer. Molecular Medicine Reports. 2019;**19**(4):3159-3167

[60] Zhou Y, Feng X, Liu YL, Ye SC, Wang H, Tan WK, et al. Down-regulation of miR-126 is associated with colorectal cancer cells proliferation, migration and invasion by targeting IRS-1 via the AKT and ERK1/2 signaling pathways. PLoS One. 2013;**8**(11):e81203

[61] Yang F, Xie YQ, Tang SQ, Wu XB, Zhu HY. miR-143 regulates proliferation and apoptosis of colorectal cancer cells and exhibits altered expression in colorectal cancer tissue. International Journal of Clinical and Experimental Medicine. 2015;**8**(9):15308-15312

[62] Sun N, Zhang L, Zhang C, Yuan Y. miR-144-3p inhibits cell proliferation of colorectal cancer cells by targeting BCL6 via inhibition of Wnt/β-catenin signaling. Cellular & Molecular Biology Letters. 2020;**25**:19

[63] Liu X, Li Y, Chen C, Li L. miR-149 regulates the proliferation and apoptosis of human colonic carcinoma cells by targeting FZD5. International Journal

of Clinical and Experimental Pathology. 2020;**13**(5):889-895

[64] Liu J, Chen Z, Xiang J, Gu X. MicroRNA-155 acts as a tumor suppressor in colorectal cancer by targeting CTHRC1 in vitro. Oncology Letters. 2018;**15**(4):5561-5568

[65] Li J, Xia L, Zhou Z, Zuo Z, Xu C, Song H, et al. MiR-186-5p upregulation inhibits proliferation, metastasis and epithelial-to-mesenchymal transition of colorectal cancer cell by targeting ZEB1. Archives of Biochemistry and Biophysics. 2018;**640**:53-60

[66] Wang B, Shen ZL, Gao ZD, Zhao G, Wang CY, Yang Y, et al. MiR-194, commonly repressed in colorectal cancer, suppresses tumor growth by regulating the MAP4K4/c-Jun/MDM2 signaling pathway. Cell Cycle. 2015;**14**(7):1046-1058

[67] Chen S, Wang Y, Su Y, Zhang L, Zhang M, Li X, et al. miR-205-5p/PTK7 axis is involved in the proliferation, migration and invasion of colorectal cancer cells. Molecular Medicine Reports. 2018;**17**(5):6253-6260

[68] Wang D, Li Y, Zhang C, Li X, Yu J. MiR-216a-3p inhibits colorectal cancer cell proliferation through direct targeting COX-2 and ALOX5. Journal of Cellular Biochemistry. 2018;**119**(2):1755-1766

[69] Zhao H, Dong T, Zhou H, Wang L, Huang A, Feng B, et al. miR-320a suppresses colorectal cancer progression by targeting Rac1. Carcinogenesis. 2014;**35**(4):886-895

[70] Mansoori B, Mohammadi A, Naghizadeh S, Gjerstorff M, Shanehbandi D, Shirjang S, et al. miR-330 suppresses EMT and induces apoptosis by downregulating HMGA2 in human colorectal cancer. Journal of Cellular Physiology. 2020;**235**(2):920-931

[71] Qu R, Hao S, Jin X, Shi G, Yu Q, Tong X, et al. MicroRNA-374b reduces the proliferation and invasion of colon cancer cells by regulation of LRH-1/Wnt signaling. Gene. 2018;**642**:354-361

[72] Han SH, Mo JS, Park WC, Chae SC. Reduced microRNA 375 in colorectal cancer upregulates metadherin-mediated signaling. World Journal of Gastroenterology. 2019;**25**(44):6495-6507

[73] Li J, Smith AR, Marquez RT, Li J, Li K, Lan L, et al. MicroRNA-383 acts as a tumor suppressor in colorectal cancer by modulating CREPT/RPRD1B expression. Molecular Carcinogenesis. 2018;**57**(10):1408-1420

[74] Wang J, Lu Y, Zeng Y, Zhang L, Ke K, Guo Y. Expression profile and biological function of miR-455-5p in colorectal carcinoma. Oncology Letters. 2019;**17**(2):2131-2140

[75] Pisano A, Griñan-Lison C, Farace C, Fiorito G, Fenu G, Jiménez G, et al. The inhibitory role of miR-486-5p on CSC phenotype has diagnostic and prognosticpotential in colorectal cancer. Cancers (Basel). 2020;**12**(11):3432-3455

[76] Wang C, Fan HQ, Zhang YW. MiR-511-5p functions as a tumor suppressor and a predictive of prognosis in colorectal cancer by directly targeting GPR116. European Review for Medical and Pharmacological Sciences. 2019;**23**(14):6119-6130

[77] Zhang W, Liao K, Liu D. MicroRNA-744-5p is downregulated in colorectal cancer and targets SEPT2 to suppress the malignant phenotype. Molecular Medicine Reports. 2021;**23**(1):54-62

[78] Yao H, Sun Q, Zhu J. miR-1271 enhances the sensitivity of colorectal cancer cells to cisplatin. Experimental and Therapeutic Medicine. 2019;**17**(6): 4363-4370

[79] Mizuno R, Kawada K, Sakai Y. The molecular basis and therapeutic potential of Let-7 MicroRNAs against colorectal cancer. Canadian Journal of Gastroenterology & Hepatology. 2018;**2018**:5769591

[80] Cruz-Gil S, Sanchez-Martinez R, GomezdeCedronM,Martin-HernandezR, Vargas T, Molina S, et al. Targeting the lipid metabolic axis ACSL/SCD in colorectal cancer progression by therapeutic miRNAs: miR-19b-1 role. Journal of Lipid Research. 2018;**59**(1):14-24

[81] Li Y, Sun Z, Liu B, Shan Y, Zhao L, Jia L. Tumor-suppressive miR-26a and miR-26b inhibit cell aggressiveness by regulating FUT4 in colorectal cancer. Cell Death & Disease. 2017;**8**(6):e2892

[82] Xu XT, Xu Q, Tong JL, Zhu MM, Nie F, Chen X, et al. MicroRNA expression profiling identifies miR-328 regulates cancer stem cell-like SP cells in colorectal cancer. British Journal of Cancer. 2012;**106**(7):1320-1330

[83] Wu N, Fesler A, Liu H, Ju J. Development of novel miR-129 mimics with enhanced efficacy to eliminate chemoresistant colon cancer stem cells. Oncotarget. 2018;**9**(10):8887-8897

[84] Wang B, Lu FY, Shi RH, Feng YD, Zhao XD, Lu ZP, et al. MiR-26b regulates 5-FU-resistance in human colorectal cancer via down-regulation of Pgp. American Journal of Cancer Research. 2018;**8**(12):2518-2527

[85] Zhu J, Xu Y, Liu S, Qiao L, Sun J, Zhao Q. MicroRNAs associated with colon cancer: New potential prognostic markers and targets for therapy. Frontiers in Bioengineering and Biotechnology. 2020;**8**:176-185

[86] Brunet Vega A, Pericay C, Moya I, Ferrer A, Dotor E, Pisa A, et al. microRNA expression profile in stage III colorectal cancer: Circulating miR-18a and miR-29a as promising biomarkers. Oncology Reports. 2013;**30**(1):320-326

[87] Kaur S, Lotsari-Salomaa JE, Seppänen-Kaijansinkko R, Peltomäki P. MicroRNA methylation in colorectal cancer. Advances in Experimental Medicine and Biology. 2016;**937**:109-122

[88] Mitchell PS, Parkin RK, Kroh EM, Fritz BR, Wyman SK, Pogosova-Agadjanyan EL, et al. Circulating microRNAs as stable blood-based markers for cancer detection. Proceedings of the National Academy of Sciences of the United States of America. 2008;**105**(30):10513-10518

[89] Ogata-Kawata H, Izumiya M, Kurioka D, Honma Y, Yamada Y, Furuta K, et al. Circulating exosomal microRNAs as biomarkers of colon cancer. PLoS One. 2014;**9**(4):e92921

[90] Fang Z, Tang J, Bai Y, Lin H, You H, Jin H, et al. Plasma levels of microRNA-24, microRNA-320a, and microRNA-423-5p are potential biomarkers for colorectal carcinoma. Journal of Experimental & Clinical Cancer Research. 2015;**34**(1):86

[91] Maierthaler M, Benner A, Hoffmeister M, Surowy H, Jansen L, Knebel P, et al. Plasma miR-122 and miR-200 family are prognostic markers in colorectal cancer. International Journal of Cancer. 2017;**140**(1):176-187

[92] Hur K, Toiyama Y, Okugawa Y, Ide S, Imaoka H, Boland CR, et al. Circulating

microRNA-203 predicts prognosis and metastasis in human colorectal cancer. Gut. 2017;**66**(4):654-665

[93] Kijima T, Hazama S, Tsunedomi R, Tanaka H, Takenouchi H, Kanekiyo S, et al. MicroRNA-6826 and -6875 in plasma are valuable non-invasive biomarkers that predict the efficacy of vaccine treatment against metastatic colorectal cancer. Oncology Reports. 2017;**37**(1):23-30

[94] Wang Q, Zhang H, Shen X, Ju S. Serum microRNA-135a-5p as an auxiliary diagnostic biomarker for colorectal cancer. Annals of Clinical Biochemistry. 2017;**54**(1):76-85

[95] Ng L, Wan TM, Man JH, Chow AK, Iyer D, Chen G, et al. Identification of serum miR-139-3p as a non-invasive biomarker for colorectal cancer. Oncotarget. 2017;**8**(16):27393-27400

[96] Miyoshi J, Toden S, Yoshida K, Toiyama Y, Alberts SR, Kusunoki M, et al. MiR-139-5p as a novel serum biomarker for recurrence and metastasis in colorectal cancer. Scientific Reports. 2017;**7**:43393

[97] He HW, Wang NN, Yi XM, Tang CP, Wang D. Low-level serum miR-24-2 is associated with the progression of colorectal cancer. Cancer Biomarkers. 2018;**21**(2):261-267

[98] Fu F, Jiang W, Zhou L, Chen Z. Circulating Exosomal miR-17-5p and miR-92a-3p predict pathologic stage and grade of colorectal cancer. Translational Oncology. 2018;**11**(2):221-232

[99] Sazanov AA, Kiselyova EV, Zakharenko AA, Romanov MN, Zaraysky MI. Plasma and saliva miR-21 expression in colorectal cancer patients. Journal of Applied Genetics. 2017;**58**(2):231-237

[100] Yan S, Jiang Y, Liang C, Cheng M, Jin C, Duan Q, et al. Exosomal miR-6803-5p as potential diagnostic and prognostic marker in colorectal cancer. Journal of Cellular Biochemistry. 2018;**119**(5):4113-4119

[101] Chen B, Xia Z, Deng YN, Yang Y, Zhang P, Zhu H, et al. Emerging microRNA biomarkers for colorectal cancer diagnosis and prognosis. Open Biology. 2019;**9**(1):180212

[102] Kjersem JB, Ikdahl T, Lingjaerde OC, Guren T, Tveit KM, Kure EH. Plasma microRNAs predicting clinical outcome in metastatic colorectal cancer patients receiving first-line oxaliplatin-based treatment. Molecular Oncology. 2014;**8**(1):59-67

[103] Rossi L, Bonmassar E, Faraoni I. Modification of miR gene expression pattern in human colon cancer cells following exposure to 5-fluorouracil in vitro. Pharmacological Research. 2007;**56**(3):248-253

[104] Li E, Ji P, Ouyang N, Zhang Y, Wang XY, Rubin DC, et al. Differential expression of miRNAs in colon cancer between African and Caucasian Americans: Implications for cancer racial health disparities. International Journal of Oncology. 2014;**45**(2):587-594

[105] Bovell LC, Shanmugam C, Putcha BD, Katkoori VR, Zhang B, Bae S, et al. The prognostic value of microRNAs varies with patient race/ethnicity and stage of colorectal cancer. Clinical Cancer Research. 2013;**19**(14):3955-3965

[106] Beg MS, Brenner AJ, Sachdev J, Borad M, Kang YK, Stoudemire J, et al. Phase I study of MRX34, a liposomal miR-34a mimic, administered twice weekly in patients with advanced solid tumors. Investigational New Drugs. 2017;**35**(2):180-188

[107] Ehrlich M. DNA methylation in cancer: Too much, but also too little. Oncogene. 2002;**21**(35):5400-5413

[108] Rodriguez J, Frigola J, Vendrell E, Risques RA, Fraga MF, Morales C, et al. Chromosomal instability correlates with genome-wide DNA demethylation in human primary colorectal cancers. Cancer Research. 2006;**66**(17):8462-9468

[109] Sheaffer KL, Elliott EN, Kaestner KH. DNA hypomethylation contributes to genomic instability and intestinal cancer initiation. Cancer Prevention Research (Philadelphia, Pa.). 2016;**9**(7):534-546

[110] Brim H, Ashktorab H. Genomics of colorectal cancer in African Americans. Next Generation, Sequencing & Applications. 2016;**3**(2):133-150

[111] Hinoue T, Weisenberger DJ, Lange CP, Shen H, Byun HM, Van Den Berg D, et al. Genome-scale analysis of aberrant DNA methylation in colorectal cancer. Genome Research. 2012;**22**(2):271-282

[112] Tse JWT, Jenkins LJ, Chionh F, Mariadason JM. Aberrant DNA methylation in colorectal cancer: What should we target? Trends Cancer. 2017;**3**(10):698-712

[113] Esteller M, Sparks A, Toyota M, Sanchez-Cespedes M, Capella G, Peinado MA, et al. Analysis of adenomatous polyposis coli promoter hypermethylation in human cancer. Cancer Research. 2000;**60**(16):4366-4371

[114] Suzuki H, Watkins DN, Jair KW, Schuebel KE, Markowitz SD, Chen WD, et al. Epigenetic inactivation of SFRP genes allows constitutive WNT signaling in colorectal cancer. Nature Genetics. 2004;**36**(4):417-422

[115] Schatoff EM, Leach BI, Dow LE. Wnt signaling and colorectal cancer. Current Colorectal Cancer Reports. 2017;**13**(2):101-110

[116] Shen L, Kondo Y, Hamilton SR, Rashid A, Issa JPJ. p14 methylation in human colon cancer is associated with microsatellite instability and wild-type p53. Gastroenterology. 2003;**124**(3):626-633

[117] Harada K, Hiraoka S, Kato J, Horii J, Fujita H, Sakaguchi K, et al. Genetic and epigenetic alterations of Ras signalling pathway in colorectal neoplasia: Analysis based on tumour clinicopathological features. British Journal of Cancer. 2007;**97**(10):1425-1431

[118] Rojas A, Meherem S, Kim YH, Washington MK, Willis JE, Markowitz SD, et al. The aberrant methylation of TSP1 suppresses TGF-beta1 activation in colorectal cancer. International Journal of Cancer. 2008;**123**(1):14-21

[119] Eads CA, Danenberg KD, Kawakami K, Saltz LB, Danenberg PV, Laird PW. CpG island hypermethylation in human colorectal tumors is not associated with DNA methyltransferase overexpression. Cancer Research. 1999;**59**(10):2302-2306

[120] Kanai Y, Ushijima S, Nakanishi Y, Sakamoto M, Hirohashi S. Mutation of the DNA methyltransferase (DNMT) 1 gene in human colorectal cancers. Cancer Letters. 2003;**192**(1):75-82

[121] Ichimura N, Shinjo K, An B, Shimizu Y, Yamao K, Ohka F, et al. Aberrant TET1 methylation closely associated with CpG Island Methylator phenotype in colorectal cancer. Cancer Prevention Research (Philadelphia, Pa.). 2015;**8**(8):702-711

[122] Turker MS. Gene silencing in mammalian cells and the spread of DNA methylation. Oncogene. 2002;**21**(35):5388-5393

[123] Xia L, Huang W, Bellani M, Seidman MM, Wu K, Fan D, et al. CHD4 has oncogenic functions in initiating and maintaining epigenetic suppression of multiple tumor suppressor genes. Cancer Cell. 2017;**31**(5):653-68.e7

[124] Wang X, Ji P, Zhang Y, LaComb JF, Tian X, Li E, et al. Aberrant DNA methylation: Implications in racial health disparity. PLoS One. 2016;**11**(4):e0153125

[125] Galon J, Costes A, Sanchez-Cabo F, Kirilovsky A, Mlecnik B, Lagorce-Pagès C, et al. Type, density, and location of immune cells within human colorectal tumors predict clinical outcome. Science. 2006;**313**(5795):1960-1964

[126] Koi M, Carethers JM. The colorectal cancer immune microenvironment and approach to immunotherapies. Future Oncology. 2017;**13**(18):1633-1647

[127] Carethers JM, Murali B, Yang B, Doctolero RT, Tajima A, Basa R, et al. Influence of race on microsatellite instability and CD8+ T cell infiltration in colon cancer. PLoS One. 2014;**9**(6):e100461

[128] Basa RC, Davies V, Li X, Murali B, Shah J, Yang B, et al. Decreased anti-tumor cytotoxic immunity among microsatellite-stable colon cancers from African Americans. PLoS One. 2016;**11**(6):e0156660

[129] Mlecnik B, Bindea G, Angell HK, Maby P, Angelova M, Tougeron D, et al. Integrative analyses of colorectal cancer show Immunoscore is a stronger predictor of patient survival than microsatellite instability. Immunity. 2016;**44**(3):698-711

[130] Yaghoubi N, Soltani A, Ghazvini K, Hassanian SM, Hashemy SI. PD-1/ PD-L1 blockade as a novel treatment for colorectal cancer. Biomedicine & Pharmacotherapy. 2019;**110**:312-318

[131] Paredes J, Zabaleta J, Garai J, Ji P, Imtiaz S, Spagnardi M, et al. Immune-related gene expression and cytokine secretion is reduced among African American colon cancer patients. Frontiers in Oncology. 2020;**10**:1498

[132] Bian S, Hou Y, Zhou X, Li X, Yong J, Wang Y, et al. Single-cell multiomics sequencing and analyses of human colorectal cancer. Science. 2018;**362**(6418):1060-1063

[133] Chen Y, Gao Y, Yi X, Zhang J, Chen Z, Wu Y. Integration of transcriptomics and metabolomics reveals the antitumor mechanism underlying Shikonin in colon cancer. Frontiers in Pharmacology. 2020;**11**:544647

[134] Ansari I, Raddatz G, Gutekunst J, Ridnik M, Cohen D, Abu-Remaileh M, et al. The microbiota programs DNA methylation to control intestinal homeostasis and inflammation. Nature Microbiology. 2020;**5**(4):610-619

[135] Paredes J, Ji P, Lacomb JF, Shroyer KR, Martello LA, Williams JL. Establishment of three novel cell lines derived from African American patients with colorectal carcinoma: A unique tool for assessing racial health disparity. International Journal of Oncology. 2018;**53**(4):1516-1528

Chapter 9

# Bayesian Random-Effects Meta-Analysis Models in Gene Expression Studies

*Uma Siangphoe*


## Abstract

Random-effects meta-analysis models are commonly applied in combining effect sizes from individual gene expression studies. However, study heterogeneity is unknown and may arise from a variation of sample quality and experimental conditions. High heterogeneity of effect sizes can reduce the statistical power of the models. In addition, classical random-effects meta-analysis models are based on a normal approximation, which may be limited to small samples and its results may be biased toward the null value. A Bayesian approach was used to avoid the approximation and the biases. We applied a sample-quality weight to adjust the study heterogeneity in the Bayesian random-effects meta-analysis model with weighted between-study variance on a sample quality indicator and illustrated the application of this approach in Alzheimer's gene expression studies.


**Keywords:** Bayesian random-effects model, meta-analysis, study heterogeneity, gene expression, sample quality weight, Alzheimer's disease

## 1. Introduction

Advances in the development of high-throughput technologies have enabled researchers to identify and quantify a large range of gene expression differences in many diseases. The number of gene expression studies has been increasing over the past decades as a result of advanced technologies. Several of them examine and address the same biological questions, even they have been implemented under different experimental conditions. Meta-analysis of gene expression data, therefore, has become a widely applied approach in combining results from multiple studies to identify common expression patterns that sometimes cannot be detected in individual studies. The meta-analytic approach has been shown to increase statistical power, accuracy, and generalizability of results [1–4]. The use of meta-analysis techniques depends on the type of response and study objectives and most analyses in microarray studies have emphasized identifying differentially expressed (DE) genes or genes that distinguish the group of samples.

Random-effects (RE) meta-analysis models are commonly applied in combining effect sizes from individual gene expression studies. However, study heterogeneity is unknown and may arise from the variation of sample quality and experimental conditions and the study heterogeneity can decrease the statistical power of the models. To maintain power, we can increase the number of studies [5] or apply an appropriate estimation method for incorporating study heterogeneity into the models. Typically, the classical RE models assume studies are independently and uniformly sampled from a population of studies. However, studies are possibly designed based on the results of previous studies. The independence assumption and an infinite population of studies may not exist. Bayesian random-effects (BRE) models have been applied to handle the uncertainty of parameters in the models. The uncertainty is incorporated through a prior distribution. A summary of evidence after the data have been observed is described by the likelihood of the models. Multiplying the prior distribution and the likelihood function will provide a posterior distribution of the parameters [6, 7].

Sample quality has a substantial impact on results of gene expression studies [8, 9]. Low heterogeneity can be found in meta-analyses containing good-quality samples, while poor-quality samples can induce high heterogeneity of effect sizes. We recently evaluated the relationships between DE and heterogeneous genes in meta-analyses of Alzheimer's gene expression data. Some overlapped DE and heterogeneous genes were detected in meta-analyses containing borderline- or poor-quality samples, while no heterogeneous genes were identified in meta-analyses containing good-quality samples [10]. The data obtained from borderline- or poor-quality samples can increase study heterogeneity and decrease the efficiency of meta-analyses [11, 12].

Small samples in gene expression studies may limit the application of classical RE models and its results may be biased toward the null or the observed value is closer to the null hypothesis than the true value. The BRE model can be used to avoid the approximation and the biases. We introduced a meta-analytic approach that included a sample-quality weight to adjust study heterogeneity in the BRE model [13]. The gene expression data therefore would include both up-weighted good-quality samples and down-weighted borderline-quality samples. Therefore, we first reviewed the classical RE models, the BRE model, and the weighted BRE model in the methods section and then illustrated an application of the methods in Alzheimer's gene expression studies. Our results are then compiled in the results section and followed by discussion and conclusions.

## 2. Methods

### 2.1 Standard random-effects model

Choi et al. [14] introduced an unbiased standardized mean difference in expression between groups for each gene $g$ [14, 15]. The meta-analytic model for combining the effect sizes is based on a two-level hierarchical model as follows:

$$\begin{aligned} y_{ig} &= \theta_{ig} + \varepsilon_{ig}, \quad \varepsilon_{ig} \sim N\left(0, \sigma_{ig}^2\right) \\ \theta_{ig} &= \beta_g + \delta_{ig}, \quad \delta_{ig} \sim N\left(0, \tau_g^2\right) \end{aligned} \tag{1}$$

where $y_{ig}$ denotes the expression for gene $g$ in study $i = 1, \dots, k$; $\theta_{ig}$ denotes the true difference in mean expression; $\sigma^2_{ig}$ denotes the within study variability displaying sampling errors conditional on $i$th study; $\beta_g$ denotes the common effects or the average measure of differential expression across individual datasets for each gene or the parameter of interest; $\delta_{ig}$ denotes the random effects; and $\tau^2_g$ denotes the between-study variability displaying the variability between studies. We estimate the parameter of the common effect using a weighted least squares estimation. We minimize the sum of squares error by differentiating with respect to $\hat{\beta}_g$ for each gene in each study, which yields.

$$\hat{\beta}_g = \frac{\sum_{i=1}^{k} w_{ig} y_{ig}}{\sum_{i=1}^{k} w_{ig}}, w_{ig} = \sigma_{ig}^{-2}. \tag{2}$$

Generally, an unbiased estimator for $\tau^2_g$ can be derived from the method of moments and the estimator can attain negative values, therefore its truncated version, or the DerSimonian-Laird (DSL) estimator, is considered instead:

$$\hat{\tau}^2_{DSL(g)} = \hat{\tau}^2_g = \max\left\{0, \frac{Q_g - (k_g - 1)}{S_{1g} - (S_{2g}/S_{1g})}\right\} \tag{3}$$

where $Q_g = \sum_{i=1}^{k} w_{ig}\left(y_{ig} - \hat{\beta}_g\right)^2$; $w_{ig} = \sigma_{ig}^{-2}$; and $S_{rg} = \sum_{i=1}^{k} w_{ig}^r$. There may be a small bias of the DSL estimator but the bias occurs where $\hat{\tau}^2_g$ is close to zero or homogeneity [16], and the bias of the DSL estimator could not be traded off by variance reductions [17, 18]. Therefore, the DSL estimator is commonly applied when fitting random-effects models for a meta-analysis [19, 20]. In this study, we estimated $\hat{\beta}_g\left(\hat{\tau}^2_g\right)$ for each gene from the microarray data with the weight $w_{ig} = \left(\hat{\sigma}^2_{ig} + \hat{\tau}^2_g\right)^{-1}$ by the generalized least squares method [14], providing the minimum variance unbiased estimator for $\beta_g$, to obtain the statistic for each gene $(z_g)$,

$$\hat{\beta}_{DSL(g)}\left(\hat{\tau}^2_{DSL(g)}\right) = \hat{\beta}_g\left(\hat{\tau}^2_g\right) = \frac{\sum_{i=1}^{k}\left(\hat{\sigma}^2_{ig} + \hat{\tau}^2_g\right)^{-1} y_{ig}}{\sum_{i=1}^{k}\left(\hat{\sigma}^2_{ig} + \hat{\tau}^2_g\right)^{-1}}, \tag{4}$$

$$Var\left[\hat{\beta}_{DSL(g)}\left(\hat{\tau}^2_{DSL(g)}\right)\right] = Var\left[\hat{\beta}_g\left(\hat{\tau}^2_g\right)\right] = \frac{1}{\sum_{i=1}^{k}\left(\hat{\sigma}^2_{ig} + \hat{\tau}^2_g\right)^{-1}}, \tag{5}$$

$$\text{such that } z_g = \frac{\hat{\beta}_{DSL(g)}\left(\hat{\tau}^2_{DSL(g)}\right)}{\sqrt{Var\left(\hat{\beta}_{DSL(g)}\left(\hat{\tau}^2_{DSL(g)}\right)\right)}} \sim N(0, 1). \tag{6}$$

The standard random-effects model currently estimates the between-study variance $\left(\tau^2_g\right)$ or the study heterogeneity using the DSL estimator.

## 3. Bayesian random-effects model (BRE)

In contrast to the classical RE model, the data and model parameters in the BRE model are considered to be random quantities [21]. We applied the BRE model to allow for the uncertainty of the between-study variance in this study. The model for gene $g$ is written as

$$\begin{aligned} y_{ig} \mid \theta_{ig} &\sim N\left(\theta_{ig}, \sigma_{ig}^2\right), \\ \theta_{ig} \mid \beta_g, \tau_g &\sim N\left(\beta_g, \tau_g^2\right), \\ \beta_g &\sim N(0, 1000), \text{and} \\ \tau_g &\sim \text{uniform } (0, 1). \end{aligned} \tag{7}$$

The kernel of the posterior distribution can be written as

$$\begin{aligned} p\left(\beta_g, \theta_{1g}, \ldots, \theta_{kg}, \tau_g^2\right) &\propto p\left(\boldsymbol{\theta}_g \mid \mathbf{y}_g, \boldsymbol{\sigma}_g^2\right) p\left(\beta_g, \tau_g^2 \mid \boldsymbol{\theta}_g\right) \\ &\propto \prod_{i=1}^{k} p\left(\theta_{ig} \mid y_{ig}, \sigma_{ig}^2\right) p\left(\theta_{ig} \mid \beta_g, \tau_g^2\right) \pi\left(\beta_g\right) \pi\left(\tau_g^2\right), \end{aligned} \tag{8}$$

where $\mathbf{y}_g = \left(y_{1g}, \ldots, y_{kg}\right)$, $\boldsymbol{\sigma}_g^2 = \left(\sigma_{1g}^2, \ldots, \sigma_{kg}^2\right)$, and $\boldsymbol{\theta}_g = (\theta_{1g}, \ldots, \theta_{kg})$ for gene $g$ in the $i$th study; $i = 1, \ldots, k$. The $\pi\left(\beta_g\right)$ and $\pi\left(\tau_g^2\right)$ are non-informative priors given as $\beta_g \sim N(0, 1000)$, and $\tau_g \sim$ uniform (0,1).

The choice of prior distributions for scale parameters can affect analysis results, particularly in small samples. With scale parameters, the distributional form and the location of the prior distributions are obtained [22]. Uniform distributions are appropriate non-informative priors for $\tau_g^2$ [6, 13].

## 4. Sample-quality weights

The quality control (QC) criteria for identifying poor-quality samples in this study were the 3′:5′ GAPDH ratio greater than 3 and/or percent of present calls less than 30% for Affymetrix arrays; and detection rate less than 30% for Illumina Bead Arrays, in addition to data visualizations [8, 23]. Poor-quality samples were excluded before data preprocessing. Furthermore, the inverse of the within-study variance is considered an optimal weight for meta-analysis. The variance of weighted mean ($\hat{\beta}_g$) is minimized when the weights are taken from the variance of the samples $y_{ig}$. A high variance gives low weights in meta-analysis [24, 25]. In our recent study, the weight intermingled with the QC indicators called as "zero-to-one weight" was most appropriate for detecting DE genes [13]. The QC indicators adjusted the within-study variance in the weighted function as:

$$w_{P6} = \left(\sigma_{ig}^{2(w_{P1})} + \hat{\tau}_g^2\right)^{-1}, \tag{9}$$

where $w_{P1} \in \{2^{-S_{ij}}, 0.01\tilde{P}_{ij}\}$ $\tilde{P}_{ij}$ denotes the percent of present calls of the $j$th sample in the $i$th study. A high value of the $P_{ij}$ weight indicates good-quality samples, providing high values of zero-to-one weights $(w_{P,ij})$ to give more weight to the expression data.

## 5. Weighted between-study variance model

We adjusted the between-study variance in the BRE model (Eq. (9)) by multiplying with an average weight over the total sample in the $i$th study for gene $g$ $\left(\overline{w}_{ig} = \sum_{j=1}^{n_{ig(a)}+n_{ig(c)}} w_{ijg} / \left(n_{ig(a)} + n_{ig(c)}\right)\right)$. The BRE weighted between-study variance model for gene $g$ is given by

$$
\begin{aligned}
y_{ig} \mid \theta_{ig} &\sim N\left(\theta_{ig}, \sigma_{ig}^2\right), \\
\theta_{ig} \mid \beta_g, \tau_g \overline{w}_{ig} &\sim N\left(\beta_g, \tau_g^2 \overline{w}_{ig}\right), \\
\beta_g &\sim N(0, 1000), \text{and} \\
\tau_g &\sim \text{uniform } (0, 1).
\end{aligned}
\tag{10}
$$

We implemented two chains each with 20,000 iterations, a 15,000 burn-in period, and a thinning of 3 in the Bayesian model, and assessed the convergence of the model using the Gelman and Rubin diagnostic [26]. The posterior mean was standardized by posterior standard deviation as the posterior distribution was symmetric and normal. We then applied a Benjamini and Hochberg (BH) procedure to control the false discovery rate (FDR) for multiple gene testing. The performance of several BRE models for unweighted and weighted data, Gibbs and Metropolis-Hastings sampling algorithms, weighted common effect, and weighted between-study variance and classical RE models for unweighted and weighted data were evaluated in simulation studies [10, 13]. The classical RE and BRE meta-analysis models were implemented using programs from *MAMA, R2jags,* and *metaDE* packages in the R programming environment [27–29].

## 6. Results

We reviewed publicly available gene expression data from the NCBI GEO database. Twelve series of RNA expression profiling in the GEO database were selected for initial review. Eligible criteria for data acquisition were as follows: (1) the datasets were publicly accessible, (2) the samples were from human brain regions, (3) the series included samples from healthy controls, (4) the datasets included phenotypic data and published manuscripts describing the data, (5) the datasets without redundant samples, and (6) the raw or normalized intensity data were defined as gene expression levels. For each study we reviewed the minimum information about a microarray experiment (MIAME) from the GEO website, including research methods and results described in the manuscripts, and data summaries of the phenotypic data. This presented study included four publicly available Alzheimer's disease (AD) gene expression datasets of post-mortem brain samples. The Gene Expression Omnibus accession numbers: GSE1297 [30], GSE5281 [31], GSE29378 [32], and GSE48350 [33] containing the gene expression and phenotypic data were included in this presented study. Some of these accession numbers (GSE5281, GSE13214, and GSE48350) include samples from multiple brain regions; we restricted our attention to only those samples acquired from hippocampus and to AD and control subjects in each dataset. The QC criteria for identifying poor-quality samples were having a 3′/5′ glyceraldehyde-3-phosphate dehydrogenase (GAPDH) ratio greater than three and/or percent of present calls

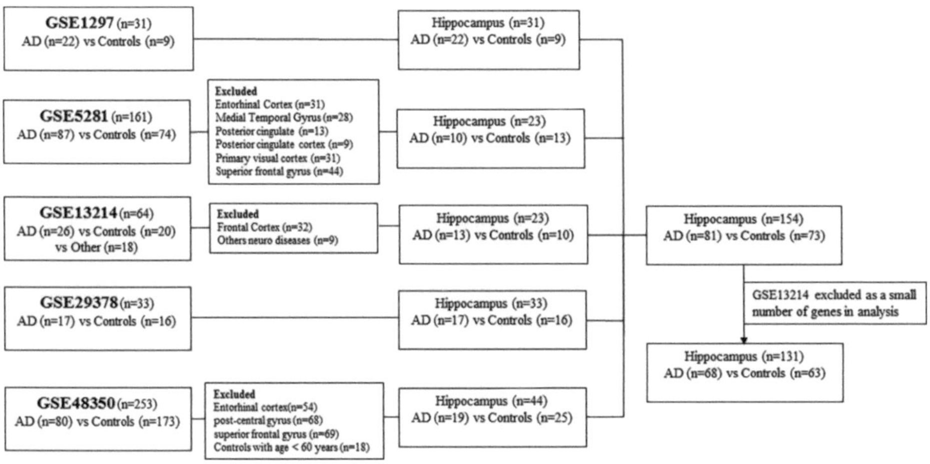


**Figure 1.**
*Study profile for meta-analysis in Alzheimer's gene expression datasets.*

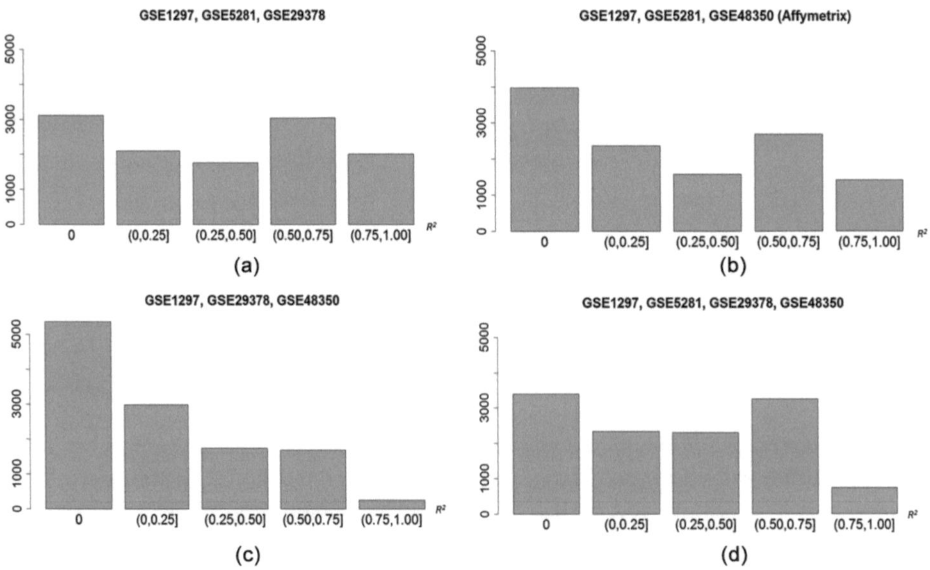


**Figure 2.**
*Barplots on strength of study heterogeneity measuring by random effects coefficient determination* ($R^2$) *in meta-analysis in Alzheimer's gene expression datasets. The* $R^2$ *of 10,000 genes were categorized into five groups. Tentatively,* $R^2$ *close to 0.25, 0.50, and 0.75 indicate low, moderate, and high heterogeneity, respectively. The y-axis presents the* $R^2$ *and the x-axis presents the number of genes in the meta-analysis.*

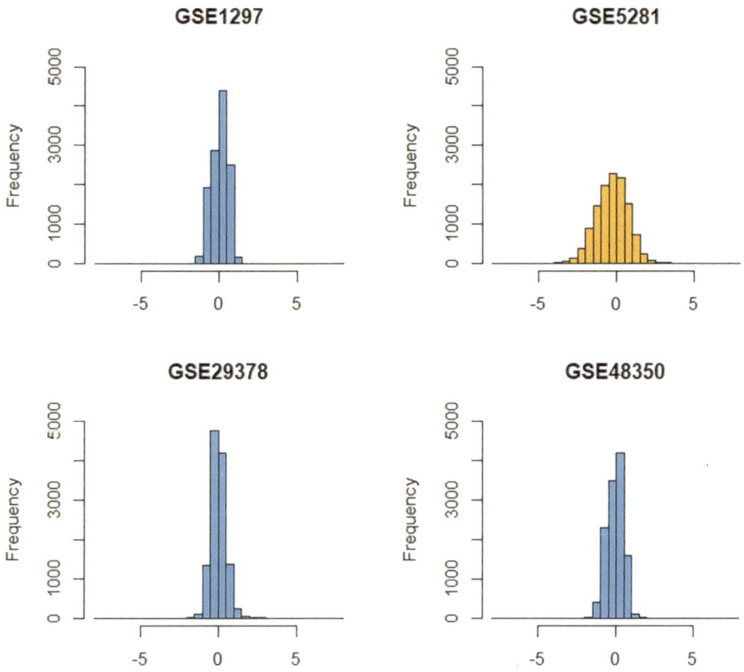


**Figure 3.**
*Distribution of unbiased standardized mean difference of gene expression (x-axis) between Alzheimer's and control groups in GSE1297, GSE5281, GSE29378, and GSE48350 datasets.*

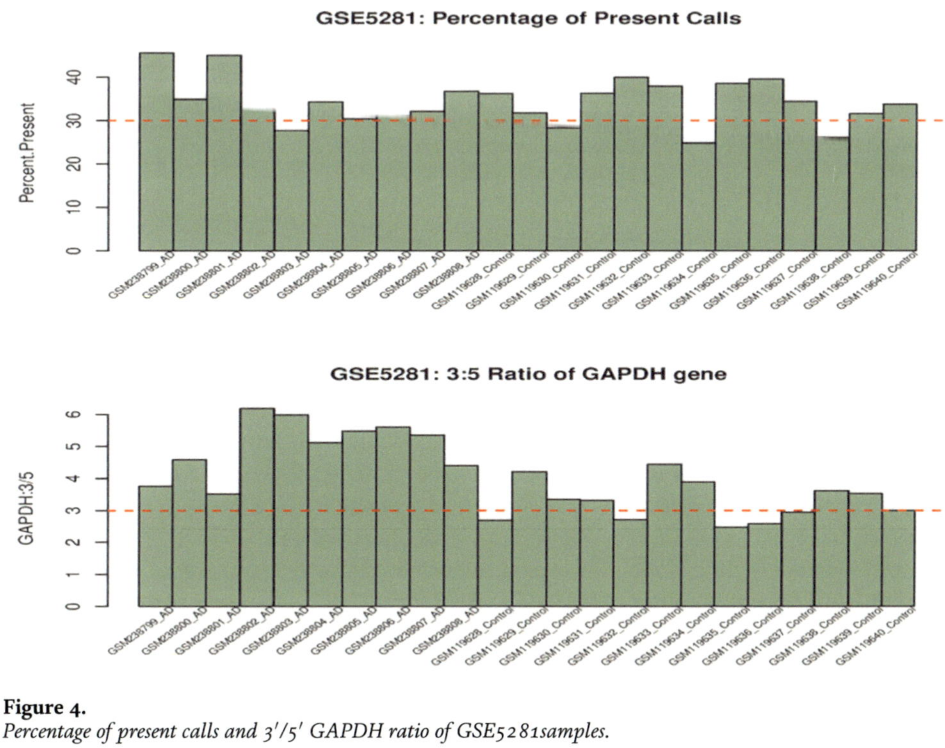


**Figure 4.**
*Percentage of present calls and 3′/5′ GAPDH ratio of GSE5281samples.*

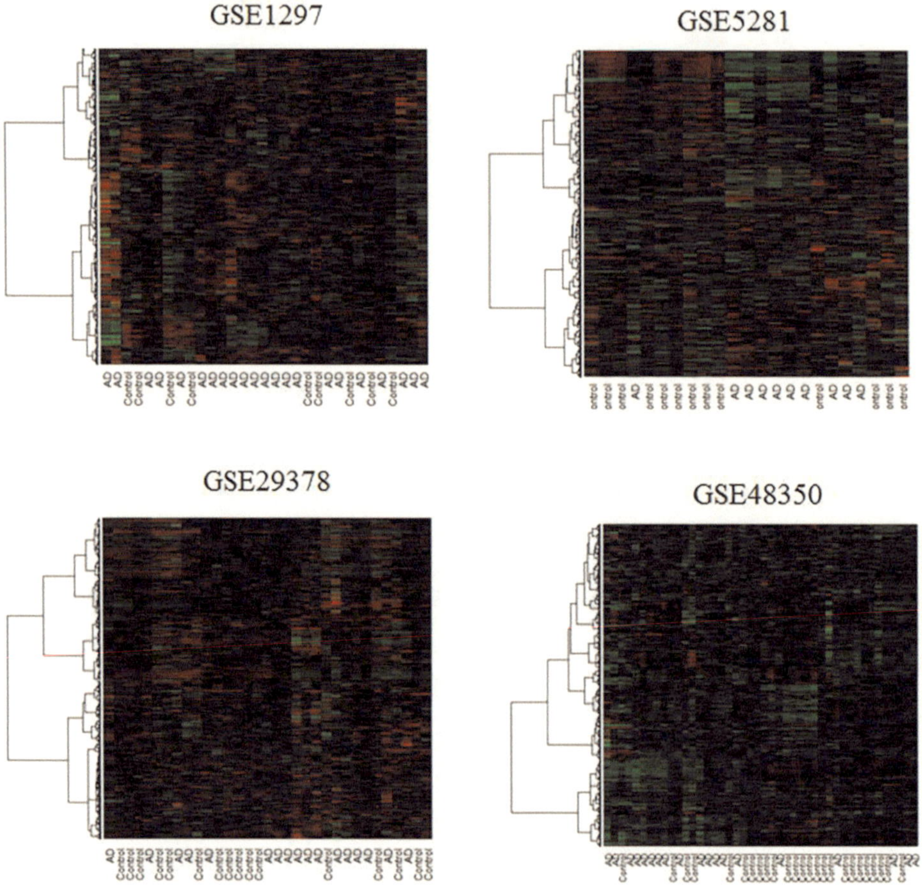


**Figure 5.**
*Heatmaps of expression patterns of 1766 differentially expressed genes in hippocampus in Alzheimer's and control samples. The differentially expressed genes were detected by the classical random-effect meta-analysis model on the metadata of four Alzheimer's gene expression datasets: GSE1297, GSE5281, GSE29378, and GSE48350.*

less than 30% [23]. We then conducted within study data preprocessing, quantile normalization, and data aggregating. Our meta-analysis was therefore performed on 12,037 target genes in 131 subjects (68 AD cases and 63 controls) from the four studies using the Affymetrix and Illumina platforms (**Figure 1**). We then considered five ways of metadata sets and primarily examined the strength of study heterogeneity ($R^2$) of each metadata object as described in [10]. The metadata A, B, D, and E had a relatively high $R^2$, while the metadata C had a relatively low $R^2$. In other words, metadata C contains homogenous data, while the remaining metadata objects may contain heterogeneous data (**Figure 2**). The distribution of unbiased standardized mean differences of gene expression in the GSE5281 dataset, which is different from the other datasets, is presented in **Figure 3**. The percent of present calls and the 3′:5′ GAPDH ratio of the heterogeneous dataset is presented in **Figure 4**.

In this meta-analysis on the metadata of the four AD gene expression datasets, 1766 DE genes were identified by the classical RE model, while 466 DE genes were identified by the weighted BRE model. Almost all the DE genes identified by

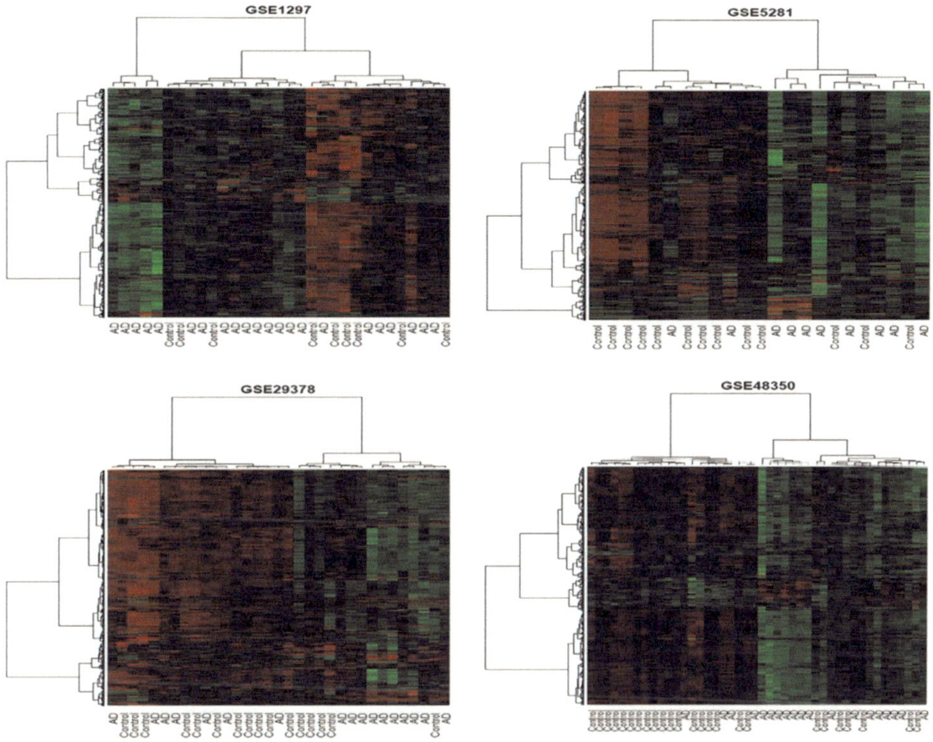


**Figure 6.**
*Heatmaps of expression patterns of 446 differentially expressed genes in hippocampus in Alzheimer's and control samples. The differentially expressed genes were detected by the classical random-effect meta-analysis model on the metadata of four Alzheimer's gene expression datasets: GSE1297, GSE5281, GSE29378, and GSE48350.*

the weighted BRE model were genes among the significant DE genes identified by the classical RE model. **Figure 5** presents the heatmap of 1766 DE up-regulated and down-regulated genes detected in the AD samples. There was no trend apparently toward more up-regulated genes or down-regulated genes on the AD samples as compared to the control samples. Meanwhile, there was a trend toward more down-regulated genes on the AD samples as compared to the control samples in the heatmap of the 466 identified DE genes (**Figure 6**). The 446 genes could potentially be down-regulated genes that may contribute to the good classification of Alzheimer's samples (**Table 1**).

We then performed gene network analysis using a publicly available web interface, GeneMANIA [34]. The 446 DE genes identified by the weighted BRE model participate in 146 significant pathways at a false discovery rate of 1%. The first-thirty highly significant pathways with more than twenty identified DE genes in each network included cellular respiration, oxidative phosphorylation, mitochondrial protein complex, inner mitochondrial membrane, protein complex, ATP metabolic process, respiratory electron transport chain, ATP synthesis coupled electron transport, electron transport chain, mitochondrial ATP synthesis coupled, electron transport, mitochondrial inner membrane, energy derivation by oxidation of organic compounds, respiratory chain complex, respirasome NADH dehydrogenase activity, NADH dehydrogenase (quinone) activity, NAD(P)H dehydrogenase (quinone) activity, mitochondrial respirasome, oxidoreductase activity, acting on NAD(P)H, quinone or

| |
|---|
| AACS, AASDHPPT, ABCA1, ACLY, ACOT7, ADAM22, ADAM23, ADARB1,AFF2, AGK, AMPH, ANGPT1, ANP32C, AP2S1, AP3B2, AP3D1, AP3M2, APBA2, APMAP, ARFGEF1, ARHGDIG, ARHGEF9, ARPC5L, ASIC2, ASNS, ASPHD1, ATAT1, ATP1A1, ATP1A3, ATP2A2, ATP2B2, ATP5B, ATP5C1, ATP5D, ATP5G1, ATP5H, ATP5L, ATP6AP1, ATP6V0B, ATP6V0E1, ATP6V1B2, ATP6V1E1, ATP6V1G2, ATP8A2, ATPIF1, ATR, ATRN, ATRNL1, ATXN7L3B, BCL2, BEX1, BEX4, BPGM, BSN, C10orf88, C12orf10, C14orf2, C16orf45, C1orf216, C2CD5, C2orf47, C5orf22, CA10, CABYR, CACNA2D3, CADPS, CALY, CAMK1, CAMK2N1, CAMKV, CAPRIN2, CCK, CDC40, CDC42EP4, CDK5, CDKN2D, CGREF1, CHGB, CHN1, CISD1, CLIP3, CLTA, CNR1, COPS3, COPS7A, COPZ2, COQ6, COX4I1, COX6C, CP, CREBBP, CRYM, CS, CUL2, CYCS, CYP4F12, DAP3, DCTN1, DDX41, DEAF1, DGUOK, DHRS11, DHRS3, DIRAS3, DLEC1, DLG2, DLGAP2, DMXL2, DNASE2, DNM1, DNM1L, DNM3, DOCK3, DOPEY1, DROSHA, DYNC1H1, DYNC1I1, ECM2, EEF1A2, EGFR, EHD3, ELF1, ELOVL4, ELOVL6, ENC1, ENO2, ENTPD2, ENTPD3, EPB41L1, EPS15, ERC2, FAM111A, FAM127A, FAM162A, FAM174B, FAM188A, FAM216A, FAM60A, FAM98A, FAR2, FGF12, FH, FHL2, FIBP, FKBP3, FMO2, FOCAD, FOXJ1, FOXO1, FRMPD4, FSD1, FXN, FYCO1, FYN, GABBR2, GABRG2, GAD, GAD2, GCC2, GLS2, GNAI2, GNG3, GNG4,GOT1, GPHN,GPI, GPRASP1, GRIA2, GRIN1, GRM1, GSTA4, GUCY1B3, GUK1v GYPC, HAGH, HARS, HERC1, HMGCR, HMP19, HN1, HNRNPUL1, HPRT1, HSPA12A, IGF1R, IMMT, IMP3, IMP4, INA, INPP5F, ITPKB, ITSN1, KAT6A, KCNN3, KCNQ2, KIAA0513, KIAA1324, KIF21B, KIFAP3, LARGE, LCMT1, LDB2, LEMD3, LGALS8, LPAR4, LPCAT4, LPIN1, LPP, LRPPRC, LRRC8B, LY6H, MAK16, MAP1A, MAP2K1, MAP2K4, MAP3K9, MAPK9, MAST3, MCF2, MCTS1, MDH1, MDH2, MICU1,MKKS, MLLT11, MOAP1, MPP1, MPPED2, MRPL15, MRPL17, MRPL35, MRPS11, MRPS17, MRPS22, MTMR11, MTSS1L, MTX2, MXI1, MYL12B, MYT1L, NAP1L2, NAP1L3, NCALD, NDN, NDRG3, NDRG4, NDUFA10, NDUFA3, NDUFA4, NDUFA8, NDUFA9, NDUFS3, NDUFS5, NDUFV2, NECAP1, NEDD8, NEFL, NEFM, NELL1, NETO2, NFIB, NIPSNAP3B, NLK, NME1, NMNAT2, NOVA1, NREP, NRGN, NRIP3, NRN1, NSF, NSG1, NUPL2, OGDHL, OPA1, ORC5, P4HTM, PAGE1, PAX6, PDCD1LG2, PEX11B, PIN1, PLCD1, PLCE1, PLCL2, PLD3, PLEC, PLEKHA4, PLK2, PLSCR4, PLXNB2, PMFBP1, PNMAL1, PNO1, PODXL2, POLB, POLRMT, POP7, PPFIA4, PPIA, PPIP5K1, PPM1H, PPME1, PPP1R13L, PPP2CA, PPP3CB, PREP, PREPL, PRKCZ, PRMT1, PRPF40A, PSD4, PSMD8, PTDSS1, PTGES2, PTPRE, PTPRR, PTRH2, PTS, PVRL3, RAB11A, RAB26, RAB27A, RAB2A, RAB6A, RAD51C, RAP1GDS1, RARS, RBFOX2, RGS17, RGS7, RHOQ, RIMBP2, RIT2, RND2, RNF123, RNF41, RNFT2, RNMT, RNPS1, RPH3A, RPP40, RPS6KC1, RUNDC3B, RWDD2A, RXRA, SCAMP2, SCG5, SCN2A, SCN3B, SDHA, SEC22A, SEC61A2, SEH1L, SEPT6, SERPINI2, SEZ6L2, SLC12A5, SLC25A11, SLC25A12, SLC25A14, SLC25A4, SLC4A1AP, SLIRP, SLITRK3, SMARCA4, SMO, SMOX, SMYD2, SNAP25, SNAP91, SNCB, SOX2, SPAG7, SPIN2A, SPINT2, SRM, SRPR, SS18L1, SSPN, STAU2, STMN2, STX6, STXBP1, SULT4A1, SUSD4, SV2B, SYDE1, SYN1, SYN2, SYNGR1, SYNJ1, SYT1, TAGLN3, TAZ, TBC1D31, TBC1D9, TBCC, TBCE, TBL1X, TBPL1, TCEA2, TCF7L2, TERF2IP, TGFBR3, THOC5, TMEM151B, TMEM160, TMEM246, TMEM59L, TMEM70, TMEM97, TNPO1, TOMM34, TOMM70A, TOR1A, TPD52, TPI1, TRAP1, TRAPPC2, TRIM37, TRIM9, TRIOBP, TSPAN13, TSPAN7, TSSC1, TUBA1B, TUBA4A, TUBB3, TUBG1, TUBG2, TXNDC9, UBE2M, UBE2S, UCHL1, UCHL3, UQCC1, UTP11L, VSNL1, WDR47, WDR7, WFDC1, XK, YWHAH, ZFP36L1, ZNF365, ZNHIT3 |

*Note: The differentially expressed genes were detected by the weighted Bayesian random-effect meta-analysis models on the metadata of four Alzheimer's gene expression datasets: GSE1297, GSE5281, GSE29378, and GSE48350.*

**Table 1.**
*List of 446 significantly differentially expressed genes in Alzheimer's gene expression datasets.*

similar compound as acceptor, respiratory chain complex I, NADH dehydrogenase complex, proton transmembrane transporter activity, aerobic respiration, presynapse, postsynapse, NADH dehydrogenase, complex assembly, oxidoreductase complex, proton-transporting two-sector ATPase complex, mitochondrial proton-transporting ATP synthase complex, ATPase-coupled cation transmembrane transporter activity, synaptic vesicle recycling, inner mitochondrial membrane organization, and cellular response to peptide. GeneMANIA overall retrieved the genes with known coexpression (51.98%), consolidated pathways (25.08%), physical interactions (27.73%), colocalization (10.79%), genetic interactions (5.79%), predicted interactions (2.65%), pathway (0.86%), and share protein domain (0.20%). More details can be found on www.genemania.org.

## 7. Discussion

In this study, we developed a meta-analytic approach for incorporating sample-quality information into the BRE meta-analysis model using an efficient weight identified by a series of simulation studies [10, 13] to adjust the study heterogeneity in the model. We illustrated the weighted Bayesian approach as compared to the classical RE model through an application in Alzheimer's gene expression studies. We have seen the results of Alzheimer's gene expression varied by the sample qualities [13]. The variation of sample quality restricted meta-analysis techniques to properly detect DE genes [35, 36]. Meanwhile, the BRE meta-analysis model allows flexibility in calculating $y_{ig}$ and its variance $\sigma_{ig}^2$ as well as study-specific adjustments [37]. We therefore can up-weight good-quality samples and down-weight borderline-quality samples in the model. This developed approach utilizes sample-quality information in the meta-analysis of high-dimensional microarray studies in detecting DE genes.

Additionally, the classical RE model tended to estimate $\tau_g^2$ as being zero and the variance of $\hat{\beta}_g$was underestimated, while the BRE meta-analysis model can allow for the uncertainty of the parameter estimates in the model. The BRE model used the marginal posterior distribution of $\tau_g^2$ for $\hat{\beta}_g$ estimation, which does not reply on the point estimate of $\tau_g^2$. The BRE model can therefore, in turn, increase the fitness of the model [38].

## 8. Conclusions

This meta-analytic approach with the sample-quality weight can increase the precision and accuracy of the Bayesian random-effects models in gene expression meta-analysis. The performance of the weighted Bayesian random-effects model may be varied depending on data feature, levels of sample quality, and adjustment of parameter estimates.

## Author details

Uma Siangphoe
Moderna, Cambridge, MA, USA

*Address all correspondence to: uma.siangphoe@gmail.com

## References

[1] Rung J, Brazma A. Reuse of public genome-wide gene expression data. Nature Reviews Genetics. 2013;**14**(2): 89-99

[2] Ramasamy A, Mondry A, Holmes CC, Altman DG. Key issues in conducting a meta-analysis of gene expression microarray datasets. PLoS Medicine. 2008;**5**(9):e184

[3] Tseng GC, Ghosh D, Feingold E. Comprehensive literature review and statistical considerations for microarray meta-analysis. Nucleic Acids Research. 2012;**40**(9):3785-3799

[4] Chang LC, Lin HM, Sibille E, Tseng GC. Meta-analysis methods for combining multiple expression profiles: Comparisons, statistical characterization and an application guideline. BMC Bioinformatics. 2013;**14**(1):1-15

[5] Borenstein M, Hedges LV, Higgins JP, Rothstein HR. Power analysis of meta-analysis. In: Introduction to Meta-analysis. The Atrium, Southern Gate, Chichester, West Sussex, United Kingdom: John Wiley & Sons Ltd; 2021. pp. 266-276

[6] Higgins J, Thompson SG, Spiegelhalter DJ. A re-evaluation of random-effects meta-analysis. Journal of the Royal Statistical Society: Series A (Statistics in Society). 2009;**172**(1): 137-159

[7] Ntzoufras I. Bayesian hierarchical models. In: Bayesian Modeling Using WinBUGS. Hoboken, New Jersey: John Wiley & Sons; 2011. pp. 305-340

[8] Draghici S. Quality control. In: Statistics and Data Analysis for Microarrays Using R and Bioconductor. 2nd ed. Boca Raton, Florida: Chapman & Hall/CRC Mathematical and Computational Biology; 2016. pp. 633-689

[9] Siangphoe U, Archer KJ. Gene expression in HIV-associated neurocognitive disorders: A meta-analysis. Journal of Acquired Immune Deficiency Syndromes. 2015;**70**(5): 479-488

[10] Siangphoe U, Archer KJ. Estimation of random effects and identifying heterogeneous genes in meta-analysis of gene expression studies. Briefings in Bioinformatics. 2017;**18**(4):602-618

[11] Cheng WC, Tsai ML, Chang CW, Huang CL, Chen CR, Shu WY, et al. Microarray meta-analysis database (M(2)DB): A uniformly pre-processed, quality controlled, and manually curated human clinical microarray database. BMC Bioinformatics. 2010; **11**(1):1-9

[12] Irizarry RA, Warren D, Spencer F, Kim IF, Biswal S, Frank BC, et al. Multiple-laboratory comparison of microarray platforms. Nature Methods. 2005;**2**(5):345-350

[13] Siangphoe U, Archer KJ, Mukhopadhyay ND. Classical and Bayesian random-effects meta-analysis models with sample quality weights in gene expression studies. BMC Bioinformatics. 2019;**20**(1):1-5

[14] Choi JK, Yu U, Kim S, et al. Combining multiple microarray studies and modeling interstudy variation. Bioinformatics. 2003;**19**(suppl 1):i84-i90

[15] Hedges L, Olkin I. Random effects models for effect sizes. In: Statistical Methods for Meta-Analysis. Orlando, FL: Academic Press; 1985. pp. 189-203

[16] DerSimonian R, Laird N. Meta-analysis in clinical trials. Controlled Clinical Trials. 1986;7(3):177-188

[17] Biggerstaff B, Tweedie R. Incorporating variability in estimates of heterogeneity in the random effects model in meta-analysis. Statistics in Medicine. 1997;**16**(7):753-768

[18] Shuster JJ. Empirical vs natural weighting in random effects meta-analysis. Statistics in Medicine. 2010; **29**(12):1259-1265

[19] Whitehead A, Whitehead J. A general parametric approach to the meta-analysis of randomized clinical trials. Statistics in Medicine. 1991;**10**(11): 1665-1677

[20] Demidenko E, Sargent J, Onega T. Random effects coefficient of determination for mixed and meta-analysis models. Communications in Statistics-theory and Methods. 2012; **41**(6):953-969

[21] Alex JS, Keith RA. Bayesian methods in meta-analysis and evidence synthesis. Statistical Methods in Medical Research. 2001;**10**(4):277-303

[22] Lambert PC, Sutton AJ, Burton PR, Abrams KR, Jones DR. How vague is vague? A simulation study of the impact of the use of vague prior distributions in MCMC using WinBUGS. Statistics in Medicine. 2005;**24**(15):2401-2428

[23] Dumur CI, Nasim S, Best AM, Archer KJ, Ladd AC, Mas VR, et al. Evaluation of quality-control criteria for microarray gene expression analysis. Clinical Chemistry. 2004;**50**(11): 1994-2002

[24] Chen D, Peace KE. Fixed-effects and random-effects in meta-analysis. In: Applied Meta-Analysis Using R. Boca Raton, Florida: CRC Press; 2013. pp. 27-52

[25] Kacker RN. Combining information from interlaboratory evaluations using a random effects model. Metrologia. 2004; **41**(3):132

[26] Gelman A, Carlin JB, Stern HS, Rubin DB. Model checking and improvement. In: Bayesian Data Analysis. Boca Raton, Florida: Chapman & Hall. CRC Texts in Statistical Science; 2004. pp. 161-197

[27] Ihnatova I. MAMA: Meta-Analysis of MicroArray. R Package Version 2.2.1. 2013

[28] Su YS, Yajima M. R2jags: Using R to Run'JAGS'. R package version 0.5-7. Available from: CRAN. R-project. org/ package= R2jags. 2015

[29] Wang X, Kang DD, Shen K, et al. An R package suite for microarray meta-analysis in quality control, differentially expressed gene analysis and pathway enrichment detection. Bioinformatics. 2012;**28**(19):2534-2536

[30] Blalock EM, Geddes JW, Chen KC, et al. Incipient Alzheimer's disease: Microarray correlation analyses reveal major transcriptional and tumor suppressor responses. Proceedings of the National Academy of Sciences. 2004; **101**(7):2173-2178

[31] Liang WS, Dunckley T, Beach TG, et al. Gene expression profiles in anatomically and functionally distinct regions of the normal aged human brain. Physiological Genomics. 2007;**28**(3): 311-322

[32] Miller JA, Woltjer RL, Goodenbour JM, et al. Genes and pathways underlying regional and cell

type changes in Alzheimer's disease. Genome Medicine. 2013;**5**(5):48

[33] Blair LJ, Nordhues BA, Hill SE, et al. Accelerated neurodegeneration through chaperone-mediated oligomerization of tau. The Journal of Clinical Investigation. 2013;**123**(10):4158-4169

[34] Warde-Farley D, Donaldson SL, Comes O, et al. The GeneMANIA prediction server: Biological network integration for gene prioritization and predicting gene function. Nucleic Acids Research. 2010;**38**:W214-W220

[35] Gentleman RC, Carey VJ, Bates DM, Bolstad B, Dettling M, Dudoit S, et al. Bioconductor: Open software development for computational biology and bioinformatics. Genome Biology. 2004;**5**(10):R80

[36] Eijssen LM, Jaillard M, Adriaens ME, Gaj S, de Groot PJ, Muller M, et al. User-friendly solutions for microarray quality control and pre-processing on ArrayAna lysis.org. Nucleic Acids Research. 2013; **41**(W1):W71-W76

[37] Demidenko E. Meta-analysis model. In: Mixed Models: Theory and Applications with R. Hoboken, New Jersey: John Wiley & Sons; 2013. pp. 247-291

[38] Bodnar O, Link A, Arendacká B, Possolo A, Elster C. Bayesian estimation in random effects meta-analysis using a non-informative prior. Statistics in Medicine. 2017;**36**(2):378-399